VISION IN THE BRAIN

VISION IN THE BRAIN

Panagiotis G. Simos

SWETS & ZEITLINGER
PUBLISHERS

LISSE   ABINGDON   EXTON (PA)   TOKYO

Library of Congress Cataloging-in-Publication Data

Applied for

ISBN 90 265 1814 5

Contents

Acknowledgements

I am indebted to Ms. Michelle Fitzgerald and Ms. Gaye Thorton for taking the time to look over an earlier version of the manuscript and for their numerous editorial suggestions. Many individuals deserve credit for taking over daily responsibilities, especially during the final stages of preparation of this monograph, first and foremost my wife Angela. My gratitude to my teacher and colleague, Dr. Andrew Papanicolaou goes beyond words: his mentoring (and patience) has made everything possible.

Panagiotis Simos

Chapter 1

Studying the Brain Mechanisms for Sensory Functions

The property that distinguishes animals (both vertebrate and invertebrate) from other living organisms is their ability to move and to respond to stimuli. The capacity of living organisms to detect changes in their immediate environment (both internal and external) is one of the main functions of the brain.[1] Incoming stimuli, when they fall within the sensitivity range of the nervous system, cause sensation in the conscious individual. By convention, researchers assume that the central nervous system mediates certain operations that produce sensation, perception, recognition, thinking, voluntary action, etc. In psychological terms the outcomes of these operations are known as *psychological constructs*. A construct can be studied experimentally only if it has been operationally defined. For instance, *color perception* can be inferred from the ability of conscious subjects to accurately *discriminate* between two visual stimuli that reflect different wavelength bands. One way to ascertain that an individual can indeed perform this discrimination is to require that he or she performs the corresponding discrimination task with better than chance accuracy.

The study of the relations between the magnitude and quality of environmental events (physical stimuli) and the intensity and quality of the sensations that they are associated with is the subject matter of *psychophysics*. In a typical psychophysical experiment the researcher records one or more behaviors and their respective parameters (e.g., speed, response accuracy) and uses them as indices of stimulus detection or recognition. These behaviors are elicited in the context of a particular experimental *task* that consists of a set of instructions that attempt to guide the subject's attention, thinking, and behavior. These instructions are delivered in the context of an experimental setting which typically consists of a set of stimuli delivered with a specific timing and require some sort of decision and response within a preset time frame. With the aid of such tasks one can determine the sensitivity limits of the sensory system under study in the form of sensory *thresholds*.

Although the study of thresholds plays a major part in psychophysical research, a more ambitious goal of psychophysisists is to uncover how sensory processes are organized to produce perception. Researchers hope to achieve this goal by systematically manipulating stimulus parameters and by changing the requirements of the experimental tasks and recording the effects of these changes on the subject's behavior. Psychophysical methods are not limited to subjects who understand verbal instructions. Modified psychophysical techniques can be applied to animals and prelinguistic children.

An important point that all students of perception should be aware of is that a psychological construct that describes a perceptual operation does not necessarily correspond to a distinct neural operation. Psychological constructs, like word recognition, are derived from the experimental study of behavior, whereas the existence of a particular neural operation is hypothesized on the basis of a specific pattern of event-related neural activity.

A more direct way to study the relation between the physical properties of external events and their perceived attributes is to examine the neural processes involved in the reception of such events. The systematic study of these relations is the main object of the field of *sensory physiology.*[2] By describing the processes through which external stimuli become coded in neural activity (i.e., they become encoded), sensory physiology attempts to bridge the gap between psychophysics and physiology. Specifically, sensory physiology studies the function of receptor organs that convert environmental events into neural signals, and the function of the ascending or afferent[3] pathways that carry these signals to the central nervous system.

The first step in the encoding of external events is sensory transduction, which can be broken down into three main stages: detection, amplification, and signaling. Detection involves the conversion of stimulus energy into another form -- mainly chemical or electrical -- and takes place in specialized receptor organs in the periphery. In vertebrate vision, the detection stage corresponds to the activation of pigment molecules in photoreceptors. In most cases, the receptors receive external input through certain accessory processes. In vision the structures that convey light to the photoreceptors and form sharp images of the external world on the layer of receptors are part of the optic apparatus of the eye (e.g., lens, iris, etc.). These structures convert the original external event (distal stimulus) into events readily accessible by the receptors (proximal stimulus). The next stage in transduction is amplification, which refers to the events that take place inside the receptor cells, usually in a cascade fashion. For instance a single activated photopigment molecule is capable of stimulating thousands of messenger molecules that serve as the final step in the conversion of neurochemical events into electrical signals. At this stage the proximal stimulus has been transformed into a code that the central nervous system can understand (or decode). In our previous example, the signaling stage corresponds to the generation of trains of neural impulses that travel along the optic nerve to reach the central structures of the visual system.

In addition, the subject matter of sensory physiology includes the structural and functional organization of the cerebral systems whose activity presumably underlies the sensation and perception of these events. In general, it attempts to uncover relations between three kinds of observations: (a) the physical characteristics of external events or stimuli, such as the intensity and frequency of sound waves, the luminance and wavelength of visual stimuli, or the amount of pressure exerted on the skin by an object; (b) the sensations or subjective experiences related to the introduction of a stimulus or to a change in one or more of its physical characteristics; and (c) changes in neural activity that correlate with the occurrence of external events and/or with sensation. The former two classes of observations are collected and analyzed using research tools borrowed from psychophysics.

Various techniques for measuring neural activity have been developed over the years and employed with the hope that they will reveal some aspects of nervous system function. These functional imaging techniques fall into two broad categories: microscopic (such as the recording of the physiological responses of individual neural units with microelectrodes), and macroscopic, which involve recording the activity displayed by large aggregates of neural elements. Techniques in the former category are invariably invasive and can only be used with non-human animals and certain patient populations as part of clinical diagnostic procedures (e.g., patients with intractable seizures who are candidates for brain surgery). Recently, a number of macroscopic recording techniques have been developed that allow the imaging of brain activity, in real time, in normal healthy individuals. Some of these techniques are completely non-invasive (for example, functional magnetic resonance imaging [fMRI], electroencephalography [EEG], and magnetoencephalography [MEG]), whereas others involve minimal risk for the individuals who are tested as subjects (i.e., positron emission tomography [PET], single-proton emission computed tomography [SPECT]). Every imaging method measures only one aspect of neural activity, i.e., neuronal signaling, blood flow, or metabolism, and attempts to identify systematic variations in this activity that may serve as indices of brain function.

The role of a particular brain area in a specific sensory process can also be established by correlating the locus of a brain lesion with sensory deficits revealed by the subject's impaired performance on one or more psychophysical tasks. The effects of experimentally induced lesions can only be studied in animals. In addition, a large amount of data has been collected from the study of the effects of brain lesions in humans that occur as a result of a neurological disease, such as stroke or tumor growth, or accidental brain injury (e.g., a bullet wound). In lesion analysis it is implicitly assumed that the impact of a particular lesion on the ability to perform a specific task is proportional to the degree to which the operations that are normally required for the execution of that task (in the intact brain) depend on that area. Despite a number of potential flaws associated with this line of reasoning, this assumption underlies the conclusions of the vast majority of published lesion studies.

At this point it will be helpful to draw a distinction between *sensory cues* and perceived *stimulus attributes* (DeYoe & Van Essen, 1988). Each sensory cue, such as orientation, contrast, and retinal disparity, may be involved in the perception of more than one stimulus attribute. For instance, contrast can be used to perceive shape as well as motion. Also each perceptual attribute, such as shape, texture, and distance, can be derived on the basis of more than one sensory cue. As we will see in Chapter 5, neurons in many areas of the visual system respond selectively to a narrow range of stimulus orientations. To conclude, however, from these observations that these areas are specialized for the analysis of shape is not justified. The same neurons could also play a role in the analysis of other stimulus attributes such as texture. In other words, the contribution of a brain area in the perception of a particular stimulus attribute cannot be inferred solely on the basis of data regarding the sensitivity of cells it contains to individual sensory cues. Thus, neurons that appear to be sensitive to a particular sensory cue are often distributed across many brain areas. Given the lack of a one-to-one correspondence between sensory cues and perceived stimulus attributes one can easily comprehend why so many functional imaging and lesion studies have failed to establish a direct link between the function of a particular brain area and the perception of a single stimulus attribute.

Pitfalls associated with functional imaging of the nervous system

The principal goal of sensory physiology, as defined in the beginning of this chapter, must be revisited. According to the revised definition, sensory physiology aims at establishing links between sensory cues, perceived stimulus attributes, and brain function. For the purposes of our discussion, we will adopt a broad definition of "brain function". According to that definition, brain function is conceptualized as a set of operations that become engaged in parallel or in series, and may involve vast numbers of neural units distributed over many brain regions (Papanicolaou, 1998). The final outcome of these operations can be an action, a thought, the feeling that a particular stimulus is familiar (recognition), or simply the cue for a behavioral response. Brain function can be inferred from a variety of measures of brain activity either at the macroscopic or the microscopic level. Modern techniques used to measure brain activity have produced an enormous amount of data in the last decade or so. Unfortunately, their capability to map brain function is often overstated, and some of the conclusions based on these data, misleading.

It should be emphasized that none of these measures can provide an accurate description of the entire set of neural operations that the brain is engaged in at any given point in time. Thus, establishing causative links between the three sets of observations may prove to be impossible with currently available techniques. For instance, recording a change in neural activity (such as an increase in the firing rate of a group of neural units in the visual cortex) in response to a change in a visual stimulus (such as the orientation of a light bar) does not necessarily imply that the observed neural response reflects the manner in which the nervous system encodes stimulus orientation. It is likely that neural units in many other brain areas show concurrent changes in activity. Unless researchers have the means to monitor the activity of the entire assembly of neural units or neuronal aggregates simultaneously and in real time, it will not be possible to establish a complete picture of neural function at a given point in time. Moreover, the entire pattern of anatomical connections between the individual components of the system, that respond in one way or another to a change in stimulus orientation, must be known. In that sense, attempts to localize a given function within a restricted area of the brain may prove entirely fruitless: every imaging method, whether it operates at the microscopic or at the macroscopic level, is at best capable of providing measures of local brain activity, rather than descriptions of localized function. In addition, establishing relations of cause and effect between neural activity and subjective experience requires that scientists have the means to monitor the contents of experience at a particular point in time in an accurate and objective manner, a premise which is in principle untenable. Thus, to further argue that an observed neural response mediates the individual's subjective perception of a corresponding stimulus change is largely a matter of conjecture.

The short-comings of modern non-invasive imaging methods can be classified into two broad categories: (a) those associated with the assumptions underlying the imaging technique itself, and (b) those related to the way that brain activity unfolds in association with an external event, a decision or behavior. With respect to the former type of problems, several properties of the measures used in non-invasive imaging render their inter-pretation ambiguous. First, in certain methods the measure of interest (the dependent variable) is relative rather than absolute. For instance, in electrophysiological research the voltage fluctuations recorded at a given scalp location are quantified in relation to voltages recorded simultaneously from other often remote scalp sites, which are pre-sumed to be "blind" to the neurophysiological event under investigation. Second, the measured quantity is affected by the properties of intervening tissue (such as conductivity in the case of EEG) and also by the geometry and spatial distribution of the neuronal ag-gregates that become active simultaneously (in both EEG and MEG research). Thus, even if the measure of interest could be recorded in an absolute way, it would still be difficult to relate the measurements directly to underlying neuronal activity. Third, cer-tain imaging methods (i.e., fMRI, PET, & SPECT) are based on subtraction techniques, wherein measurements are conducted under a minimum of two experimental events (or conditions). The final outcome of a study depends upon a number of apparently arbitrary decisions such as the choice of the "control" task, the statistical method on which the contrast between the two conditions is based etc. In summary, the correspondence be-tween the dependent variable in each technique and underlying brain activity is by no means a direct one. Measured events bear a complex, non-linear, and often obscure rela-tion to actual neurophysiological events.

The advantage of invasive approaches (such as the use of intracranial recordings) to mapping brain activity is their ability to provide direct information regarding the function of individual neural units in response to experimental manipulations. They suffer, how-ever, from sampling limitations: in most cases activity from a single nerve cell is moni-tored one at a time. Unless combined with microlesion procedures they do not provide sufficient information to ascertain whether the neurons in a particular area that change their activity in response to an external event are indeed essential the performing the function associated with this event.

The results from invasive as well as non-invasive procedures are often difficult to in-terpret because, normally, the brain is engaged in many tasks at any given time. Further, each of these tasks can be mediated by a number of alternative mechanisms. Moreover, it is difficult to determine the onset and the temporal course of the engagement of a par-ticular mechanism, even in the controlled setting of the laboratory. Scientists do not know how each of the functional units of the brain operates -- for instance some units may operate in a "digital", all-or-none fashion, others as "analogue" devices, like the postsynaptic membrane of the nerve cell, and others in some other yet unspecified man-ner. Finally, there is insufficient information regarding how these units are anatomically and functionally interconnected. Perhaps the most commonly used strategy for extracting activity that is specific to a particular task is averaging, whereby the task of interest is repeated a sufficient number of times. It is assumed that brain activity associated with other concurrently executed operations, which are not time-locked to the task of interest is averaged out, leaving only activity corresponding to the operations necessary for the task of interest. This may well be so in many circumstances. Yet even when the process of averaging is successful in procuring a reliable representation of task-specific brain activity, it still does not provide a description of the *mechanisms* that mediate the execu-tion of the task of interest. Once again, description of the mechanism entails specification of the algorithm that governs the mode of activation of the brain structures and not only visualization of what structures are activated and to what degree. This is precisely what functional images are unable to procure at present.

Yet despite this limitation, currently available functional imaging methods are making and will be making unique contributions. First of all, a variety of non-invasive techniques are now available, which allow researchers to appreciate the pattern of brain activity in the intact human brain with higher precision, under experimental situations that afford a greater degree of ecological validity than ever before. The functional images that can now be recorded, though they do not reveal mechanisms, do give an outline, however fuzzy and incomplete, of functional systems that mediate, somehow, certain cognitive operations which, in turn, are loosely associated with particular types of tasks. Second, one could expect that with the integration of information obtained with many different methodologies (e.g., non-invasive imaging, lesion studies in humans and animals, single cell recordings, in vitro neurochemistry, etc.) regarding the operational mode of smaller functional units and their interconnections, it will become possible to achieve realistic descriptions of the mechanisms themselves. Finally, some of the problems associated with the interpretation of measures of brain function can be overcome by careful study design. This typically involves performing measurements under two or more different experimental events (or conditions), which may consist of different stimuli, tasks, and levels or categories within a given task. With those considerations in mind we will proceed to review specific applications of two widely used imaging techniques, event-related potentials and evoked magnetic fields.

In Chapter 1 of this book, a very brief overview is presented of the general principles of brain organization with emphasis on the function of the brain's basic elements, the nerve cells. Then, in Chapters 3-6, the organization of the visual system is discussed in detail. The experimental findings reviewed in these chapters are relevant to both the anatomy and the physiology of that system. In most cases the presented data were obtained with invasive methods, electrophysiological and histological, employed in experiments on non-human primates. When available, data from human populations are also included for comparison. An effort is made to highlight relations between observed anatomical and physiological characteristics of each neural system and perceptual phenomena as studied with psychophysical methods. The latter were obtained using non-invasive procedures from neurologically intact human subjects, from patients who had suffered various types of central nervous system damage, and from animals using special training procedures. The final chapters (7-9) are devoted to a review of the development of structural and functional elements of the visual system. The focus of the discussion will be on the basic principles of maturation of the neurophysiological processes that sustain sensation and perception in the visual modality. Whenever applicable, questions pertaining to the relative contribution of genetic and environmental factors are discussed.

Terminology and general principles of brain anatomy and organization

A good way to understand how the central nervous system is organized anatomically is to visualize the spinal cord and brain (cerebrum) as part of a long axis, known as the neuraxis. This imaginary line is vertically oriented along the spinal cord and brainstem (medulla, pons, and midbrain). At the roof of the midbrain this axis bends forward by approximately 90° as it extends from the frontal pole to the occipital pole of the brain. For the purposes of this book, the term *rostral* means in the direction of the cerebrum, while *caudal* means toward the occipital pole. The dorso-ventral axis is oriented perpendicular to the rostro-caudal axis: *dorsal* indicates toward the upper surface of the cerebrum (when the body is in the upright position), whereas *ventral* means in the direction of the lower surface of the cerebrum (see Figure 1). In order to identify internal CNS structures, the brain is cut along three planes (see Figure 2). Horizontal sections are cut parallel to the plane defined by the rostro-caudal axis and the line that passes through the center of the eyes. The sagittal plane also passes through the rostro-caudal axis but is perpendicular to the horizontal plane. Parasagittal sections are parallel to the sagittal plane off the midline. Finally, coronal sections run on planes perpendicular to both the sagittal and the horizontal planes.

The elementary functional units of the nervous system are the nerve cells or neurons. Neuronal aggregates in the nervous system, that can be distinguished from the surrounding neural tissue using anatomical or physiological criteria, are given various names such as: *ganglion* (in the peripheral nervous system), *nucleus, lamina, cortex,* and *body.* Neurons are interconnected through nerve fibers (or axons). A bundle of nerve fibers that run between neuronal aggregates can be called: *tract,* fasci*culus, brachium, column, lemniscus, commissure,* or *capsule.*

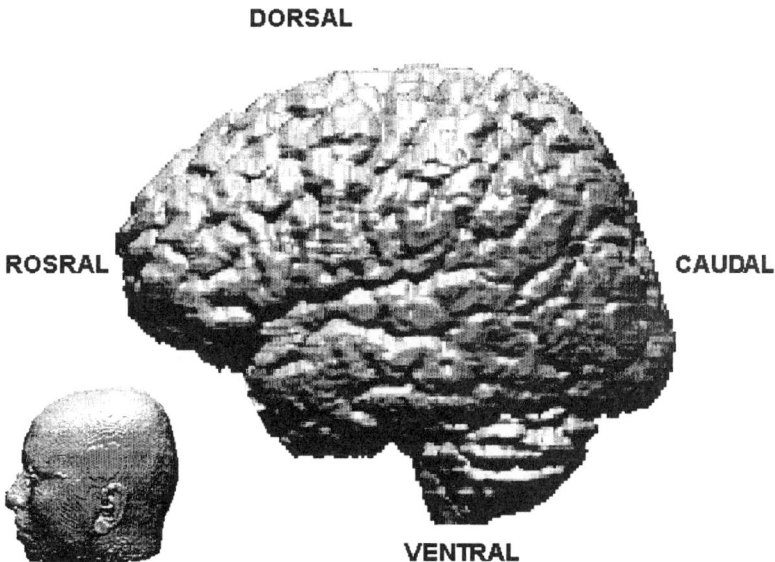

FIGURE 1.
Lateral view of the left hemisphere of a human brain.

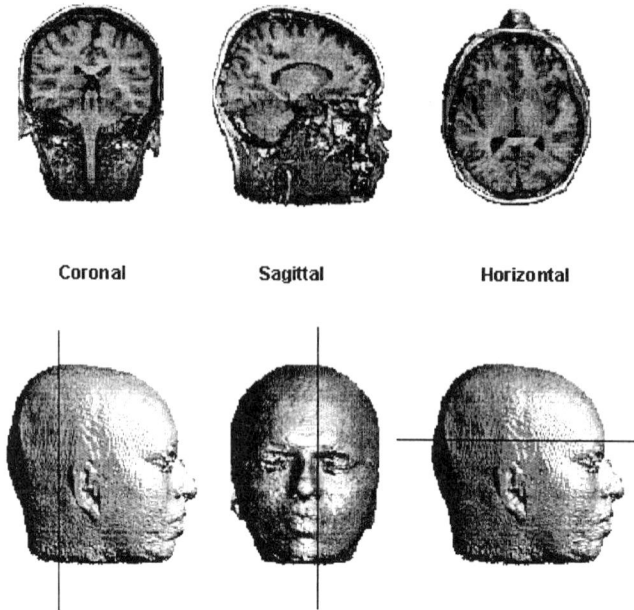

FIGURE 2.
Magnetic resonance imaging scans of a human brain, indicating the three planes used for cross sections.

Often this book will refer to experimental studies that examine the functional properties of individual neurons, i.e., changes in neuronal activity in response to a change in one or more of the characteristics of a visual stimulus. This strategy does not rest on the assumption that the functional properties of single neural units represent accurately the function of the larger neuronal population to which they belong. Although this may be true in certain cases, neurons form densely interconnected local networks so that the behavior of the network as a whole cannot be predicted from the functional properties of its constituent elements. In those cases, a mathematical representation of the responses of several neurons sampled from a given neuronal population (i.e., the population response) may serve as a more accurate index of the function of the neuronal aggregate under investigation. The development of new techniques for recording the activity of several neurons at the same time has enabled researchers to study the functional relations between neurons that are part of local networks. These investigations can provide unique insights into the *dynamic* behavior of neuronal networks, which, due to the complexity of the interactions between the individual neurons, cannot be predicted from isolated records of single unit activity.

In the periphery, specialized cells detect changes in physical energy coming from the organism's environment, and respond with changes in their electrical state. Receptors convey these changes to sensory neurons with which they are linked using chemical messengers. In the visual system, the bodies of sensory neurons are located on the interior surface of the eye bulb. Finally, long fibers (afferent fibers) carry the sensory cells' electrical responses to the central nervous system.

By convention, the central nervous system is divided into three major parts: the brainstem, diencephalon, and telencephalon (see Figure 3). The brainstem is further divided into the medulla (the part located nearest to the spinal cord), the pons, and the midbrain (or mesencephalon). The medulla contains the first relay station of afferent fibers that carry signals related to audition (cochlear nuclei) and somatic sensation (dorsal column nuclei) into the central nervous system. The midbrain contains the tectum, which mainly consists of four nuclei, the superior colliculi (one on each side of the midline) and the inferior colliculi. The superior colliculi receive visual input and are involved in the organization and control of fast eye movements. The inferior colliculi serve as a relay station for auditory fibers. The part of the diencephalon that this book is concerned with is the thalamus and specifically, a pair of thalamic nuclei that serve as the final relay station for visual input known as the dorsolateral geniculate nuclei.

The telencephalon is divided into two hemispheres by the longitudinal fissure that runs on the sagittal plane. Each cerebral hemisphere is covered by a mantle of gray matter (known as the cerebral cortex or neocortex[4]), which consists of the bodies, and processes of neurons, and supporting glial cells. The neocortical cells are arranged in six layers oriented parallel to the surface of the brain. The external surface of the hemispheres is convoluted and displays an intricate pattern of gyri and sulci. The cerebral cortex can be further subdivided on the basis of gross anatomical, functional, and histological criteria. With respect to gross anatomical features, each hemisphere can be divided into four lobes: frontal, temporal, parietal, and occipital. The frontal lobe occupies the area rostral to the central sulcus (see Figures 3 and 4). The occipital lobe occupies the posterior (caudal) part of the cerebrum.

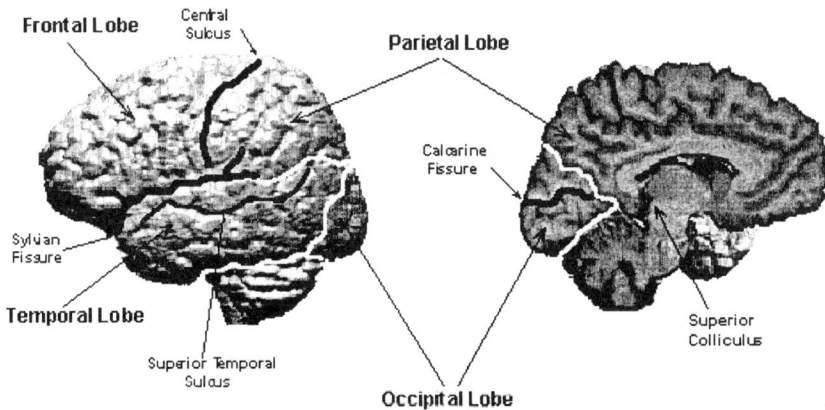

FIGURE 3.
Lateral (left-hand image) and parasagittal (right-hand image) view of the left hemisphere of a human brain indicating the location of main sulci and fissures.

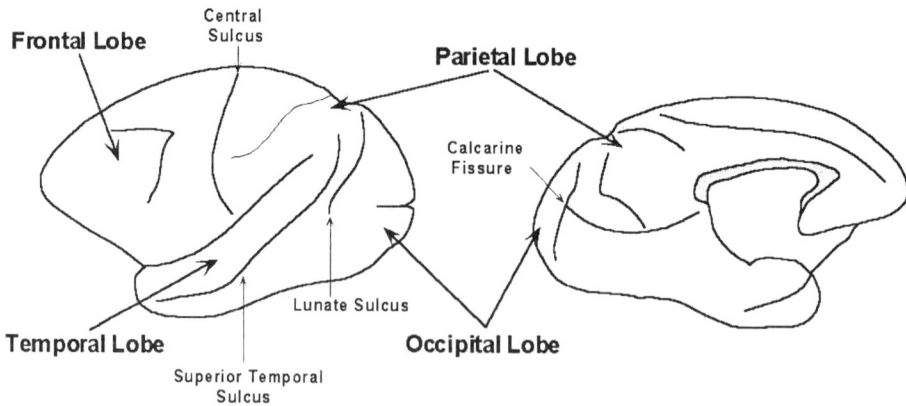

FIGURE 4.
(left) Lateral and (right) parasagittal view of the brain of the macaque monkey (Macaca mulatta).

The temporal lobe lies beneath the Sylvian fissure and extends into the ventral aspect of the hemispheres. The parietal lobe lies posterior to the central sulcus, and borders with the temporal and occipital lobes.

In addition, the neocortex can be subdivided according to the afferent modality served by each region. The occipital, temporal, and parietal lobes contain the primary receptive areas for the three major afferent modalities: vision, audition, and somatic sensation, respectively. As a rule, these areas serve as the first relay stations in the cerebral cortex for neural signals that originate in the periphery. The primary receptive area for auditory inputs lies in the transverse gyri of Heschl located on the lateral bank of the Sylvian fissure. For vision, the primary receptive area lies on the banks of the calcarine fissure, which is located in the medial aspect of each occipital lobe. Somatic inputs are first input into the cerebral cortex in three separate regions on the posterior bank of the central sulcus. In addition, the three lobes contain association areas which are believed to be involved in carrying out neural operations that lead to the encoding and recognition of complex attributes of external stimuli such as the shape, texture, and form of objects, complex patterns of visual motion, spatial relations, and complex sounds, including speech. The posterior part of the frontal lobes is occupied by areas involved in the programming and initiation of movement and in the production of speech (motor and premotor areas). The anterior part of the frontal lobes is believed to be involved in regulating complex forms of interactive social behavior, and in the planning of goal-directed action.

Using histological criteria, the cerebral cortex has been subdivided into a number of cytoarchitectonic regions. The most widely used cytoarchitectonic map is the one constructed by Brodmann a century ago. To a certain extent, similar maps have been found in other mammals. A fundamental assumption in neuroanatomy is that all mammals have evolved from a common ancestor and that all brains have derived from the same basic structure through a series of phylogenetic alterations. This notion does not imply that the brains of different species do not differ. Certain areas are shared by all primate species (e.g., "visual" areas V1, V2, and the middle temporal area; Kaas, 1993). A certain amount of variability in the precise location of borders in the architecture of neocortex is expected not only between species, but also between members of the same species (see Figure 4). Although establishing a correspondence between primary receptive areas in different species on the basis of histological criteria is a relatively easy task, things are much more complicated with respect to association areas. In that case researchers rely on other sources of information such as comparisons of connection patterns in different regions of the brain, the sensitivity of individual neuronal units to various external events, and the effects of focal lesions on behavior (Felleman & Van Essen, 1991).

A key property of all the modality-specific neocortical areas is topographical organization. Neural signals that carry inputs from the periphery arrive in these areas in an orderly manner, so that adjacent cortical loci receive input from adjacent points in the corresponding array of specialized receptors. In the visual and the somatosensory system the receptor organs have a true topographical correspondence with points in each sensory field (i.e., the visual field and the body surface and musculature, respectively). In the auditory modality, the afferent map contains a point by point representation of the array of hair cells in the inner ear that transduce the mechanical energy of sound into electrical impulses. In contrast to the other two systems, the receptor organs in the auditory system are arranged according to the sound frequency to which they are maximally sensitive, rather than forming a topographical map of auditory space (i.e., of the locations of the sound sources). A property shared by all three modalities is that each is served by more than one cytoarchitectonically distinct neocortical areas. Each of these areas contains a more or less precise afferent map. The different regions within each modality are organized in a parallel and hierarchical fashion. Thus, a single region serves as the primary recipient of visual and auditory input. The remaining areas are believed to be involved in increasingly more complex analyses of the afferent input. In the somatosensory system, the afferent input arrives initially in three neocortical areas. Topographical maps of the body present in additional areas reveal increasingly complex analysis operations.

Finally, within each neocortical area, neurons form anatomically and functionally distinct patches or columns. Neurons in each column often show preference for a narrow range of sensory cues, such as the orientation of a light bar, the eye (left or right) that has been stimulated, etc. Intrinsic connections between the neurons that belong to a single column may play a role in shaping or sharpening the preferences of individual units. Connections between columns may be involved in integrating inputs that stimulate adjacent yet separate portions of the receptor surface (see Chapter 5).

NOTES

[1] The other two major functions of the brain -- the organization and initiation of behavior and the maintenance of a constant internal environment (homeostasis) -- are outside of the focus of this book. In general, one may say that the brain controls the main adaptive functions of the organism. For this reason terms such as stimulus (and stimulation) will be used to refer to events that occur outside of the body: light reflected from surrounding surfaces, vibration of air particles produced by rapid movement of external objects, pres-

sure and thermal energy applied on the skin, etc. Stimuli generated within the body, such as the level of glucose in the blood, are part of homeostatic feedback mechanisms, or serve as signs of bodily functions, such as visceral pain. Certain stimuli produced inside the body are involved directly in the control of the organism's reactions to its environment, such as kinesthetic input that signals limb position and movement. In some cases internally generated input interacts with sensory processes. One such example is the proprioceptive input that signals eye position and movement. This is thought to be important for perceiving a stable visual world during movement of the eyes.

[2] The decision to restrict the scope of this book to sensory processes does not reflect endorsement of an empiricist viewpoint. It is acknowledged that the contents of consciousness may be determined to some extent by sources other than immediate sensory input. Moreover, the properties of sensations are determined largely by certain anatomical and physiological characteristics of the corresponding sensory systems. A basic assumption that is held throughout this book, however, is that sensory organs are the only means that the brain possesses in order to gain information regarding the environment that surrounds living organisms.

[3] In this book we prefer the term afferent to the term sensory, because the latter is implicitly associated with the notion that neural signals transmitted by these neurons necessarily lead to conscious sensation. "Afferent" is a more general term that includes inputs to the central nervous system that do not produce a conscious sensation, such as signals triggered by external stimuli that are too weak to be perceived, and others triggered by stimuli of sufficient magnitude that never capture the subject's attention.

[4] The term *cortex* is also used to refer to the gray matter that covers the external surface of other brain structures. Thus, there is the gray matter of the hippocampus, which, on the basis of evolutionary criteria, is a more primitive structure (hence called archeocortex), the cerebellar cortex, etc.

Chapter 2

Neuronal Excitability and Synaptic Transmission

Introduction to neurons and their function

Neurons are similar to the other cells of the body in that they are surrounded by a semi-permeable membrane that encloses the nucleus and the cytoplasm. The hypothesis that neurons are self-contained cells that serve as elementary functional units in the nervous system was first introduced by the pioneer neuroanatomist Ramon y Cajal at the turn of the last century. This notion became known as the *neuron doctrine*[1] or neuron theory. Conclusive evidence demonstrating that there is no cytoplasmic continuity between neighboring neurons was made available relatively recently with the help of the electron microscope. This device is capable of magnifying images by up to 100,000 times compared to the maximum magnification of 1-2,000 times of the conventional light microscope used by the early anatomists. Nerve cells, like the one shown in Figure 5, vary widely in size and shape, but share a number of special features that differentiate them from other body cells. Each neuron consists of the soma that contains the nucleus, a number of thin filaments that branch like a tree (the dendrite), and a (usually) long process (the axon), which often has several branches, known as collaterals. Axons vary in length from a few millimeters to more than one meter in certain kinds of cells found in humans and other large animals.

A major step in our understanding of neuronal function was the discovery that the excitation state of one neuron can be conveyed onto neighboring nerve cells in the form of electrical impulses. Similarly, the level of electrical activity in peripheral nervous system neurons (or sensory neurons) that are attached to peripheral receptor organs changes as a function of the excitation state of these receptors. The axons of some of these neurons form long bundles (peripheral nerves). The activity that propagates along nerve axons is in the form of a series of brief, repeated electrical impulses, which are called *spikes*, for their unique morphology when viewed on an oscilloscope screen (see Figure 6). It has been ascertained that, for a given type of neuron, the amplitude of individual spikes is relatively constant.

The neuron's response to inputs from other neurons is expressed as a train of spikes. The temporal pattern of these discharges, as opposed to the magnitude of individual electrical impulses, is known to vary systematically with certain stimulus parameters, which may be different from one cell to the next.

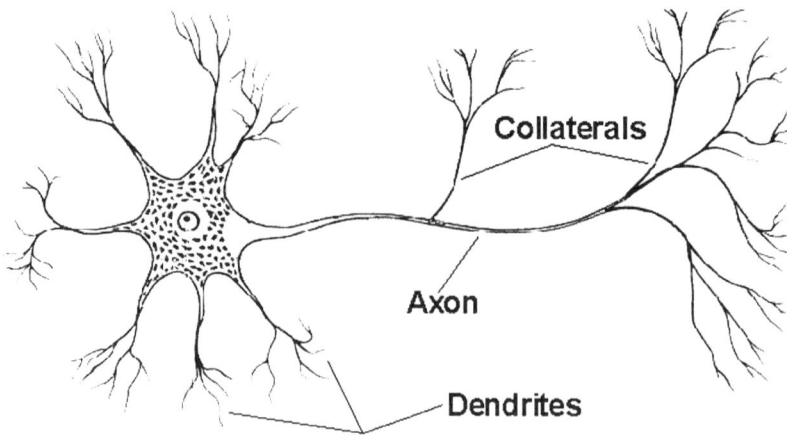

FIGURE 5.
Schematic representation of a nerve cell. Several dendritic branches that form the dendritic arbor, and two main axon colaterals are shown. Note the branching of the axon in several axonal terminals. These terminals can form contacts with a large number of postsynaptic neurons. Also, each of the terminals can form several synapses on a single postsynaptic dendritic process.

The most commonly used measure of a neuron's response is spike frequency or firing rate (i.e., the number of spikes per unit time). More recently, other measures have been explored, which, apparently, provide valuable information regarding the details of neuronal signaling (see Chapter 4). In this book the term "neural signal" refers to activity along a nerve axon in the form of a train of electrical impulses. Although this type of activity plays a crucial role in the communication between neurons, it does not directly propagate from one neuron to another, but rather requires the action of a synaptic mechanism (see below). Axonal impulses recorded from one or more cells at the same time serves as an index or sign of neuronal function.

As shown in Figure 7, a typical CNS neuron receives chemical or electrical signals from other neurons at the dendrites and soma, and generates electrical signals that travel along the axon terminal in the form of electrical impulses. Axon terminals transmit these signals via synaptic contacts to the dendrites (*axo-dendritic* synapses) or the soma of other cells (*axo-somatic* contacts). Although these two types of contacts are the most common, synaptic transmission that occurs between two axons (*axo-axonic* synapses) or between the processes that belong to two dendrites (*dendro-dendritic* contacts) are also found in the mammalian CNS.

Axo-dendritic synapses can be made on the dendritic branch directly (shaft synapses), or on specialized structures that emerge from the dendrites (dendritic spines: spine synapses). Axo-somatic synapses are often inhibitory. In which case, the synaptic effect consists of an increase in the conductance of the postsynaptic membrane to chloride ions which counteracts depolarizing currents that are conducted electrotonically (i.e., passively) from more distal parts of the cell (shunting inhibition).

(a)   (b)

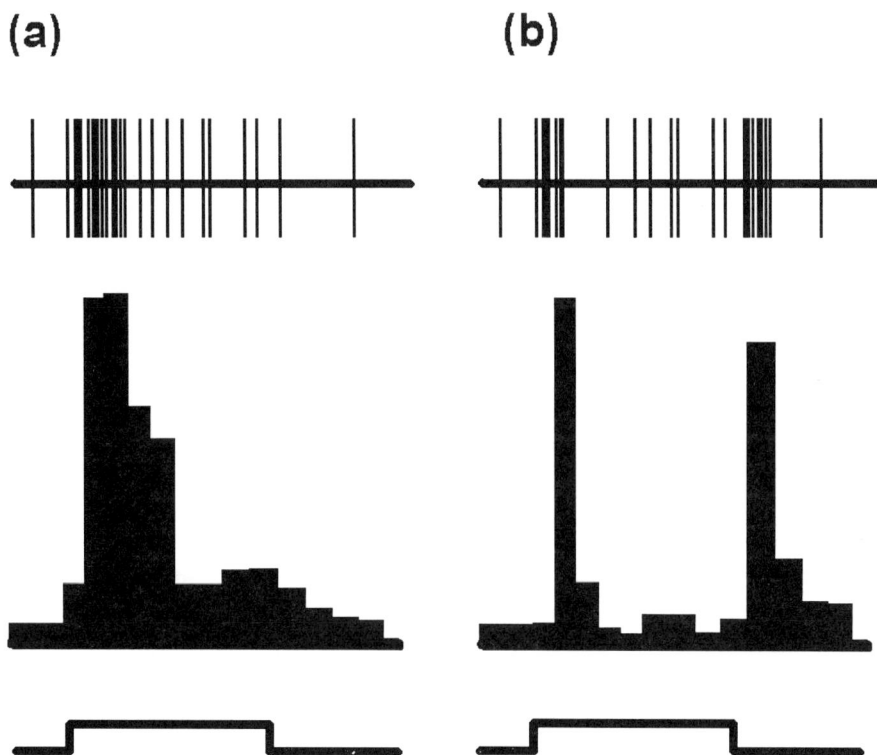

FIGURE 6.

(top) Extracellular recordings from a microelectrode placed near the axons of two second order sensory neurons (retinal ganglion cells). The two cells can be distinguished on the basis of the temporal pattern of their response to a small light spot. The neuron on the left shows a sustained response that persists throughout the duration of the stimulus, although firing rate progressively decreases during that time. The neuron on the right shows a transient response: its firing rate increases abruptly, shortly after stimulus onset, then rapidly falls until the light spot disappears, at which point a second rapid increase in firing rate is observed. It is important to realize that the response of a particular neuron may vary considerably from one repetition of the stimulus to the next. In order to obtain a more stable picture of the temporal firing pattern that characterizes each neuron, it is customary to average responses across many repeated presentations of the same stimulus. The resulting graphs are known as peristimulus time histograms or poststimulus time histograms *(middle portion of the figure)*. The average number of spikes recorded in each time window is shown on the y axis. In this case, the length of the time window or bin is 50 ms. Notice that the cells continue to fire long after the cessation of the stimulus. These discharges are produced spontaneously by most cells in the CNS and are commonly referred to as "baseline" activity.

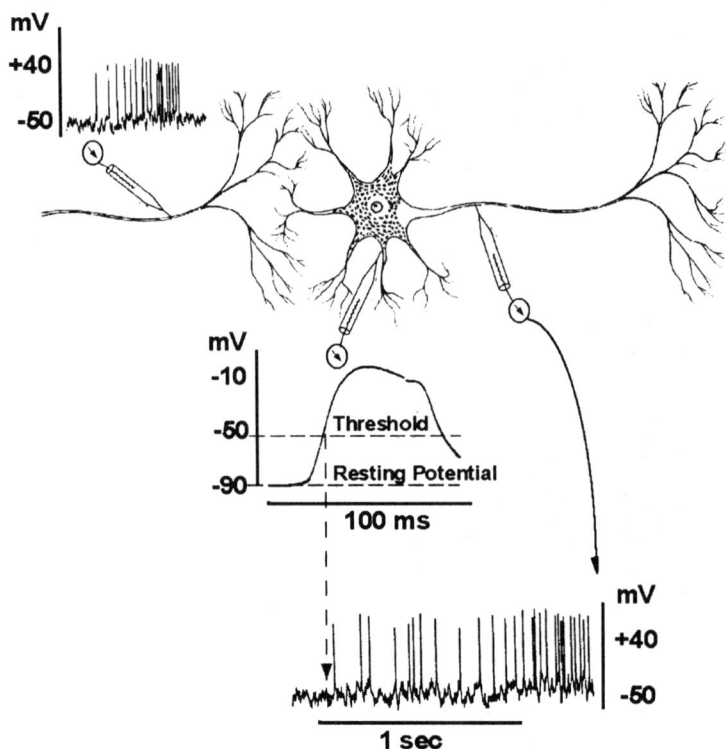

FIGURE 7.

Intracellular recordings from two neurons connected via synaptic contacts. Top: recordings from the axon of the presynaptic cell reveal a train of spikes. The voltage scale is in milli-volts and corresponds to the voltage (or potential) difference between the cytoplasmic side of the membrane and the extracellular space. When the action potentials reach the axon terminals small amounts of the chemical messenger that is produced in that cell are re-leased. A certain number of these molecules are bound on specific recognition sites in post-synaptic receptors. In this case the receptors are linked to cation channels (i.e., channels permeable to positively charged ions like sodium and potassium). The membrane potential (middle graph) is at the resting state, meaning that the cytoplasmic side of the membrane is around 90 mV more negative than the extracellular space. The potential difference between the two sides of the membrane forces cations to move toward the intracellular space. Ini-tially, only sodium ions flow into the cell through the newly opened (gated) channels. This brings the membrane potential to more positive values (depolarization). As explained in more detail later in the text, this graph displays an Excitatory Postsynaptic Potential. The new potential difference "spreads" passively along the membrane, past the cell body, to-ward the axon hillock. As a result of hundreds of these synaptic events, which often take place on different parts of the dendritic tree, the membrane potential at the hillock reaches a critical value (threshold); action potentials are generated (lower graph).

Many inhibitory synapses share common structural characteristics and can be classified as Type II synapses. The latter are characterized by a narrow synaptic cleft. In addition, membrane segments that contain high-density regions in electron microscope photographs are smaller both in the presynaptic and the postsynaptic membrane. On the other hand, spine synapses are usually excitatory and can be classified as Type I synapses. They display a wider synaptic cleft and larger high-density regions in both the presynaptic and the postsynaptic membrane, and are associated with round synaptic vesicles. Examples of such synapses are those formed by geniculocortical afferents on spiny stellate cells in layer IVC of the visual cortex in mammals (more details are given in Chapter 5).

In most cases, communication between cells is carried via chemical "messengers". Small quantities of these compounds, which are known as *neurotransmitters* are released from the axon terminals into the narrow junction that separates the axonal (or presynaptic) membrane from the membrane of the target cell (the postsynaptic membrane). Specialized structures embedded in the latter react chemically with the transmitter molecules and produce a change in the electrical properties of the postsynaptic membrane which can, under certain conditions, generate electrical impulses that travel along the axon. Neurotransmitters comprise one class of chemical agents that are broadly referred to as *messengers*. The other class consists of molecules that are released or activated inside the postsynaptic cell as a result of synaptic action and operate as second messengers or mediators in a chain of biochemical events to change certain aspects of postsynaptic cell function. To summarize, we have, thus far, mentioned two important principles of neuronal organization. First, under normal circumstances neural signals propagate in one direction, from the dendrites to the axon. Second, there is no cytoplasmic continuity between adjacent neurons. A junction (gap) known as the synaptic cleft, separates the axon terminal from the presynaptic membrane.

Neurons can be classified in several ways. One classification scheme focuses on their shape, and also on the number and pattern of their processes. *Bipolar* cells have a single main dendrite that branches at the end and a single axonal process. One of the most common functions of bipolar cells is to carry signals from the periphery to the central nervous system. The majority of neurons in the central nervous system are *multipolar* cells. They possess a single axon and several dendritic branches that emerge from different sites on the cell body. The rich dendritic tree of multipolar cells can accommodate a large number of synaptic contacts. For example, some cells with extensive and dense dendritic trees may receive up to 150,000 synaptic contacts.

Nerve cells fall into three main classes depending on their role in neuronal circuitry: afferent neurons, motor neurons, and interneurons. Afferent neurons carry signals generated in the periphery in specialized types of nerve cells known as receptor cells. *Motor* neurons carry signals from the central nervous system -- such as the motor area of the neocortex and the spinal cord -- to the effector systems (muscles or glands). Finally, *interneurons* transmit signals between central nervous system neurons. Local interneurons have short axons and connect cells located in the same region of the CNS. Examples of local interneurons are cells in the retina of the eye that establish local circuits with receptor cells and sensory neurons.

Nerve cells owe their signaling capability to their electric properties. The fact that they are enclosed in a semi-permeable membrane and that the intracellular fluid has a different concentration of certain ions (primarily sodium, potassium, and chloride) compared to the extracellular fluid, produces a separation of charges which in turn, leads to a difference in electrical potential between the two surfaces of the membrane. This voltage is known as the *resting potential* of the cell membrane (see Figure 7). Neurons (and muscle fibers), unlike other body cells, are excitable: they can change their membrane potential in response to changes in their environment. These include mechanical stimulation (which triggers the sensations of touch, pressure, some forms of pain, audition, and body orientation and balance), stimulation by light particles (which triggers visual sensations), and chemical stimulation (which is responsible for olfactory and gustatory sensations). The sensitivity of specialized nerve cells to these stimuli forms the basis for the mechanisms that supply the nervous system with afferent input.

As mentioned earlier, neurons communicate with each other mainly by chemical messengers. Released from the axon terminal, these agents can produce transient changes in the electrical state of the postsynaptic membrane. The membrane owes this capacity to certain paths or channels that traverse the lipid bilayer and allow charged particles (ions) to cross the membrane. Ion channels can be classified according to the mechanism that determines their state (open or closed). *Gated* channels increase the likelihood of being open in response to specific conditions. *Voltage-gated* channels respond to a change in the electrical state of the membrane produced by other means. *Transmitter-gated* channels have specialized sites that bind chemical messenger molecules. This, in turn, regulates the probability that the ion channel will remain in the open state. Finally, *mechanically-gated* channels respond to mechanical changes imposed on the membrane (such as pressure and stretch). *Non-gated* channels, on the other hand, remain open continuously. Ion flow through non-gated channels is regulated by osmotic and electrical factors.

Chemical messengers released from an axonal terminal can affect the electrical state of the postsynaptic membrane in three ways: (1) They can reduce the difference in electrical potential that exists between the two surfaces of the postsynaptic membrane (*depolarization*), thereby increasing the likelihood that the postsynaptic cell will generate an action potential. Externally triggered changes in membrane potential toward more positive values are also known as Excitatory Postsynaptic Potentials (EPSPs). (2) They can increase the difference in electrical potential between the two sides of the postsynaptic membrane (*hyperpolarization*), thereby reducing the likelihood of firing. The effect of the former process is considered excitatory, whereas the effect of the latter process is inhibitory. Changes in membrane potential toward more negative values are also called Inhibitory Postsynaptic Potentials (IPSPs). (3) The binding of the chemical messenger can increase the conductive properties of the postsynaptic membrane and consequently reduce the capacity of excitatory inputs that arrive concurrently, to trigger action potentials. This process is known as silent or *shunting inhibition*. There is a tendency for inhibitory synapses to be located on the soma of the postsynaptic cell, whereas excitatory synapses are more often encountered on the distal portion of the dendritic branches. In many cases excitatory synapses are made on specialized dendritic processes, known as spines.

Initially, the electrical events that occur in the postsynaptic membrane consist of graded potentials (either depolarizing or hyperpolarizing). Graded potentials are changes in the membrane potential produced by a transient change in its conductivity. This change is caused by the binding of chemical messenger molecules onto a sufficient number of receptor sites embedded in the membrane. One characteristic of these potentials is that they are associated with electrical currents that move passively along the membrane. In the vast majority of neurons (excluding photoreceptors and certain types of neurons in the retina), graded potentials can trigger action potentials. The type of chemical messenger released by the axon terminal and the type of receptor that binds the messenger molecules are the two main factors that determine the nature of electrical events in the postsynaptic membrane. The variety of receptors that are sensitive to the same chemical messenger accounts for the observation that a given messenger can produce opposite effects in different postsynaptic cells (i.e., depolarization in one cell and hyperpolarization in another cell).

When several graded potentials occur in close temporal contiguity they can sum up and jointly reduce the potential difference between the two sides of the membrane. The depolarization then spreads via passively conducted electrical currents toward the initial portion of the axon (known as the axon hillock). When the membrane potential at the axon hillock reaches a certain value (or firing threshold that is usually 5 to 15 mV more positive than the resting membrane potential) it can trigger action potentials that are very brief in duration (usually in the order of 1 to 2 ms; see Figure 7). Action potentials, which are also known as spikes for their unique morphology when viewed on an oscilloscope screen, can be transported over large distances along the axon without any significant loss of amplitude. Action potentials are generated in an all-or-none fashion: their amplitude is the same in a particular axon regardless of the amplitude and duration of the graded potential that triggered them, as long as the membrane potential has reached the firing threshold.

The primary process involved in the generation of the action potential is the activation of voltage-gated sodium channels that depolarizes the membrane. In certain neurons, other voltage-gated channels may also be involved (for instance calcium channels). Often, excitatory (depolarizing) synaptic events are counteracted through the action of nearby synapses by increasing the conductivity of the membrane to chloride ions (*shunting inhibition*). The temporal pattern of electrical impulses displayed by a given neuron depends upon the relative concentration of sodium and potassium channels in the axon hillock. Thus, a given neuron may produce a different discharge pattern than another neuron in response to the same net depolarizing current.

Mechanisms of synaptic transmission

The arrival of action potentials at the nerve ending triggers the release of neurotransmitter molecules into the synaptic cleft. CNS neurons release more than two dozen different substances that act as neurotransmitters. Several other molecules, namely hormones (e.g., insulin) and neuromodulators (e.g. prostaglandins) can also affect the functional state of neurons. Although any given cell may be influenced by more than one of these molecules, as a rule each cell produces and releases only one type of neurotransmitter. The binding of neurotransmitter molecules onto specific sites embedded into the postsynaptic membrane, leads to changes in the electrical state of the membrane. This is achieved through changes in membrane permeability that are controlled by ion channels. A better understanding of the varieties and functions of ion channels will help elucidate the mechanisms of synaptic activity. It is believed that ion channels transport ions by means of a "selectivity filter": a narrow region that removes waters of hydration from ions and forms a weak chemical bond with them. This model can explain both the speed and the relative selectivity with which ion channels operate. The flow of ions through the channel is passive, i.e., it does not require energy expenditure once the channel opens. Further, the rate of ionic flow is determined by three factors: 1) the channel conductance for a particular ion species, 2) the concentration gradient of that ion across the membrane and, 3) the electrochemical driving force that is exerted upon the ions on either side of the membrane.

As previously mentioned, there are two main families of channels, *gated* and *non-gated*. Gated channels usually remain closed when the membrane is at the resting state. Non-gated channels on the other hand, stay open continuously and are, to a large extent, responsible for the maintenance of the resting membrane potential. The state of an ion channel (open or closed) is determined by conformational changes in the protein molecule that forms the channel.

Channel gating is controlled by one of three main mechanisms:

- Through the binding of a chemical messenger molecule to a specific subunit of the receptor-ion channel complex (*directly-gated* channels). These receptors are known as *ionophoric* or *ionotropic* receptors. This is the predominant method of synaptic transmission mediated by acetylocholine (Ach) via nicotinic receptors, glycine, $GABA_A$[2] and glutamate (via the kainate, AMPA[3], and NMDA[4] receptors).

- Through the binding of a neurotransmitter onto a receptor which subsequently activates ion channels via a *second messenger* system (in*directly-gated* channels). These receptors are known as metabotropic receptors. Initially, the change in the conformational state of the receptor mobilizes a transducing protein (G-protein) that lies on the cytoplasmic side of the membrane, which in turn activates an enzyme that mediates the production of the second messenger. This agent then enables the activation of the molecules of a secondary effector substance (usually a protein kinase), which mediates the production or transport of other protein molecules, or simply triggers a change in their conformational state. These proteins are involved in a particular function of the cell, which is the ultimate target of the second messenger system. This train of events can take place in the postsynaptic as well as in the presynaptic cell and the resulting changes can either be transient or long lasting. Typically the synaptic effects mediated by second messenger systems have a delayed onset and often last longer than the synaptic effects of directly-gated channels. Second messenger systems mediate the intracellular action of norepinephrine, dopamine, serotonin, acetylocholine via muscarinic receptors, and glutamate via the quisqualate receptor (formerly referred to as quisqualate-B).

- Through a change in the membrane potential. Among the known types of receptors, only the NMDA class of glutamate receptors is controlled by both a voltage change and by neurotransmitter binding.

A characteristic feature of virtually all CNS synapses is the convergence of several inputs (made by one or more presynaptic neurons) onto more than one synaptic zone in the postsynaptic cell. In most CNS synapses, the EPSP produced by a single synaptic input is typically not sufficient to trigger an action potential. Thus, many EPSPs produced in close temporal synchrony at a number of synaptic sites are required to bring the membrane potential of the postsynaptic cell to threshold. The factors that determine the capacity of the membrane for integrating multiple synaptic events that occur in close temporal and spatial proximity are described in the following section.

Integration of postsynaptic potentials in central synapses

The ability of the postsynaptic membrane to integrate synaptic currents is determined mainly by two passive membrane properties: the time constant (τ) and the length constant (λ). The time constant is an index of the capacity of the membrane to integrate successive EPSPs induced at the same region on the membrane. It reflects the rate at which the membrane potential changes following the application of a synaptic current, and is proportional to the electrical properties of the membrane, namely resistance and membrane capacitance. The time constant is also an index of the time it takes for the membrane to return to its resting state. The longer the time constant, the slower the return of the membrane potential to its original value and the higher the likelihood for temporal summation of synaptic potentials that arrive successively at the same postsynaptic site. The value of the time constant ranges between 1 and 20 ms across different types of neurons.

The length constant reflects the distance away from the site of the application of a synaptic current at which the EPSP is reduced to 37% of its peak value. The magnitude of λ depends primarily upon the resistive properties of a particular postsynaptic segment. Again, its value is determined by passive membrane properties, namely the ratio of membrane resistance to axial resistance. The latter depends largely upon the length of the cytoplasmic segment that the synaptic current has to traverse. Membrane resistance, on the other hand, is predominantly affected by the thickness of the membrane insulation. Axial resistance increases linearly as a function of the distance from the site of the origin of the synaptic event, although membrane resistance remains constant at any distance from the site of the application of the synaptic current. As a consequence, the change in membrane potential produced by a single synaptic current becomes increasingly smaller as one moves away from the site of its generation. Thus, the better the conducting properties of the cytoplasmic core and the better the insulation of the postsynaptic membrane, the longer the distance will be over which a given EPSP will maintain a substantial proportion of its original value. In other words, a membrane with a large length constant is better equipped to integrate synaptic inputs that arrive simultaneously at neighboring sites on its surface. Typical values of λ range between 0.1 and 1 mm for different types of neurons.

Inhibitory synaptic action in the CNS

In most central neurons, current that flows through transmitter-gated chloride channels is primarily responsible for inhibitory effects. GABA, which is the most widespread inhibitory transmitter in the brain, operates through ionotropic receptors ($GABA_A$) directly linked to a chloride channel. This channel is blocked by picrotoxin and kept open by the GABA agonist muscimol, while the receptor itself is blocked by the GABA antagonist bicuculine. A second type of GABA receptor ($GABA_B$) is found primarily near presynaptic terminals and appears to be involved in the control of neurotransmitter release. When this metabotropic receptor, which is linked indirectly via a G-protein to a calcium channel, is activated through axo-axonic contacts, it can cause a reduction in the intracellular calcium concentration and a corresponding decrease in transmitter release from that terminal. This type of presynaptic modulation is common among first and second order afferent neurons (i.e., neurons located early in a sensory pathways). Presynaptic inhibition can regulate the spread of excitatory synaptic effects within an array of cells and is the principal mechanism in *lateral* or *surround inhibition*. In other instances, presynaptic inhibition can control the spatial extent of excitatory inputs carried by the different collateral branches of a single terminal tree.

Excitatory transmission pathways in the CNS

Receptors associated with directly-gated channels
Ionotropic receptor systems contain a recognition site, which binds the transmitter molecules, and an ion channel, both made of a single protein macromolecule. When the recognition site becomes occupied by a neurotransmitter molecule, it triggers a conformational change in the receptor protein, which causes the ion channel to open. The most abundant class of ionotropic receptors in the CNS are those that bind glutamate, which is the major excitatory neurotransmitter in the brain. There are three types of ionotropic glutamate receptors, each named after the synthetic amino acid for which the receptor shows the highest chemical affinity: kainate, AMPA (formerly known as quisqualate-A) and NMDA receptors. The two former receptor types bind the synthetic glutamate agonists, kainate and AMPA, respectively. The kainate receptor also binds quisqualate, while neither the kainate nor the AMPA receptors bind NMDA. The kainate and AMPA receptors are linked to a low conductance ion channel that is permeable to both sodium and potassium.

The third type of ionotropic glutamate receptor shows a unique property: the activity of the NMDA-receptor complex is controlled by the binding of chemical messengers, but is also regulated by a voltage dependent-mechanism (see Figure 8). This feature renders the NMDA receptor suitable for playing a central role in neuronal-activity dependent changes in the CNS. The NMDA receptor itself shows high affinity for the glutamate agonists aspartate (which is naturally present in the brain) and for the synthetic agonist NMDA, while another natural amino-acid, glycine, binds on a specific recognition site and exerts a modulatory effect on the receptor's activity. Glycine generally increases the affinity of the NMDA receptor for glutamate.

The NMDA receptor is blocked by the synthetic antagonist APV[5] (or AP5). In addition, the NMDA-receptor gated channel is blocked by the hallucinogenic drug phencyclidine (PCP) and also by ketamine. The receptor controls a channel that is permeable to positive ions (primarily to calcium but also to sodium and potassium). When the membrane is at the resting state, the pore of this channel is occluded by magnesium ions located near the extracellular side of the channel protein. When the membrane becomes depolarized through the action of non-NMDA receptors (often other types of glutamate receptors), the magnesium block is removed and, when glutamate and glycine are also present, the channel allows calcium, sodium, and potassium to flow through.

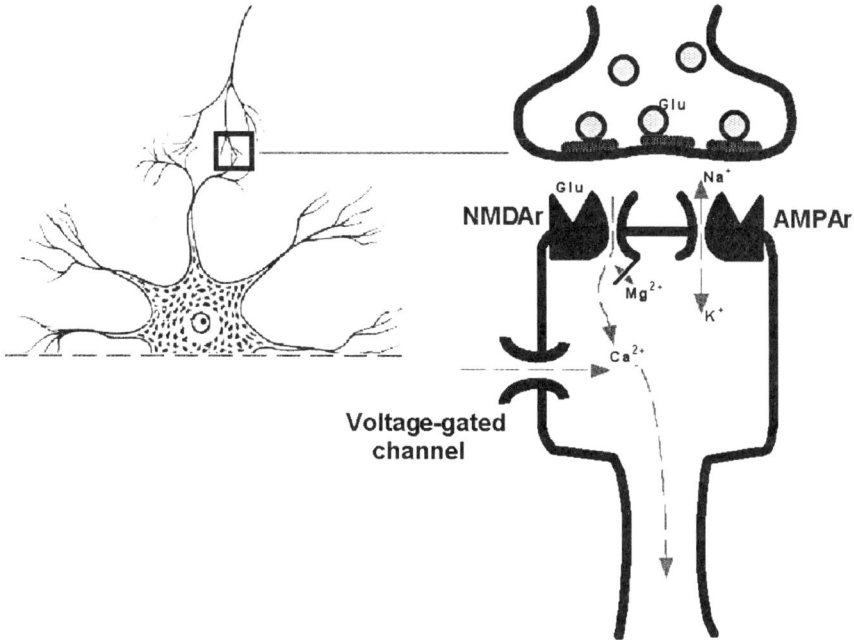

FIGURE 8.
Schematic representation of a CNS synapse that utilizes glutamate (Glu). The neurotransmitter is stored in vesicles in the presynaptic terminal (button). Active zones, involved in the docking of vesicles on the terminal membrane and in transmitter release, are shown as dark bars. An NMDA glutamate receptor (NMDAr) and an AMPA receptor (AMPAr) are shown. Both receptors are linked to a cation channel. The one linked to the NMDA receptor is usually held closed by a magnesium (Mg^{2+}) ion. Binding of glutamate on the AMPA receptor depolarizes the membrane (through the influx of positively charged sodium ions), and removes the magnesium block, allowing calcium ions (Ca^{2+}) to flow into the cell.

Calcium activates calcium-dependent protein kinases (such as the calcium/calmodulin kinase, the protein kinase A, etc.). These are capable of inducing long-term changes in the postsynaptic cells, including long-term enhancement or suppression of synaptic efficiency. It has been recently established that some of these changes are induced through the activation of *immediate early genes*, which, in turn, control the synthesis, or regulation of specific proteins, including various enzymes and neurotrophic factors. More details on the significance of these intracellular processes will be discussed in subsequent chapters.

Second messenger systems
Synaptic transmission in second messenger systems has two main components: a receptor that recognizes and binds neurotransmitter molecules released from the presynaptic terminal, and an effector system that controls a gated ion channel indirectly. The sequence of events that is described below is common to all second messenger pathways. Binding of a transmitter molecule on a specialized receptor protein causes the receptor to activate a G-

protein located on the intracellular surface of the cell membrane which, in turn, operates as a catalyst in the production of the second messenger (including cAMP,[6] diacylglycerol, or arachidonic acid).

The next stage involves an interaction between the second messenger and a protein kinase (such as the cAMP-dependent protein kinase, the protein kinase C, or lipoxygenase, respectively). The result is an activated protein kinase, which in turn, phosphorylates a cytoplasmic protein. The latter can be an enzyme, a cytoskeletal protein, or a channel protein. Depending upon the kind and location of the target protein, the phosphorylation can lead to:

- a change in the affinity of the enzyme;
- a change in the enzyme's position within the cell, which, consequently, will either suppress or enhance its action;
- a conformational change on the cytoskeletal protein;
- a conformational change leading to the opening or closing of an ion channel. For instance, the closing of non-gated potassium channels, which reduces outward leakage currents, can increase the excitability of the neuron or even depolarize the membrane directly;
- a conformational change in the receptor's own molecule that reduces its capacity to bind neurotransmitter molecules which, in turn, alters its responsiveness to feature synaptic events;
- a conformational change in one or more of the subunits of a different type of receptor. In this way second messengers can modulate transmission through other transmitter systems;
- the triggering of protein synthesis (or simply an increase in the rate of production) in the postsynaptic cell. This action is mediated by interactions with enzymes involved in the production of these proteins and/or with changes gene expression.

Second-messenger systems are also involved in the regulation of synaptic action via mechanisms that operate in the presynaptic cell. Presynaptic regulation can occur through: 1) autoreceptors that bind the cell's own neurotransmitter and 2) receptors that bind a different neurotransmitter or chemical messenger. In every case the chain of events in the presynaptic cell targets the concentration of intracellular calcium ions. At least one phenomenon of long-term synaptic modification, the persistence of synaptic potentiation known as LTP, has been linked to one such mechanism. Although the induction of LTP is mediated by processes that take place in the postsynaptic cell, the long-term maintenance of synaptic enhancement appears to depend, at least in part, upon processes that take place presynaptically, namely the sustained increase in glutamate release from presynaptic terminals. This results from the accumulation of calcium ions in the presynaptic terminal, sufficient enough to sustain a higher rate of vesicle mobilization and transmitter release long after the presynaptic fiber has returned to spontaneous discharge levels. It is likely that a chemical agent released from the postsynaptic cell diffuses to the presynaptic cell, and activates one or more second messengers, which, in turn, facilitate transmitter release via a calcium-dependent mechanism. Phenomena of long-term synaptic modification are believed to play a role in *synaptic plasticity* (i.e., persisting changes in synaptic efficiency that depend upon preceding levels of neuronal activity). As explained in more detail in Part II of this book, synaptic plasticity plays an important role in mediating the effects of sensory experience in both the mature and developing organism.

NOTES

[1] This notion should not be confused with a more recent use of the term to argue that the activity of individual neurons is sufficient to signal the presence of a particular complex visual pattern or object (see Barlow, 1972).

[2] Gamma aminobutyric acid.

[3] Alpha-amino-3-hydroxy-5-methyl-4-isoxazole proprionic acid.

[4] N-methyl-D-aspartate.

[5] 2-amino-5-phosphonovalerate.

[6] Cyclic adenosine monophosphate.

Chapter 3

Vision

Visual sensations are produced when light reaches the internal surface of the eye. Every visual scene contains information that can be described along several independent dimensions or sensory cues. The role of the visual system is to extract biologically relevant cues that allow the observer to initiate an appropriate behavioral response in a particular situation. This task is complicated by several visual factors such as changes in ambient illumination, as well as non-visual factors, such as movement of the eyes, the head, and the body of the observer, displacement of the objects themselves, context variations, etc.

Light is a form of electromagnetic radiation and can be described in two ways: as particles of energy, and as waves that travel through a medium. As an electromagnetic wave, light can be described in terms of *intensity* and *frequency* or *wavelength* (which is the inverse of frequency). The primate eye is sensitive to a restricted range of wavelengths, usually between 400 and 700 nanometers (nm). Spatial variations in the intensity of light reflected from various surfaces create the sensation of local brightness transitions or *brightness contrasts*[1] (see Figure 9). Although the perceived amount of brightness in a particular portion of the visual field is lawfully related to the physical properties of reflected light (i.e., its intensity or luminance), it is also affected by the luminance of neighboring surfaces. Thus, the same gray square appears brighter when surrounded by a darker surface than when surrounded by a brighter surface. An example of a context effect in the perception of brightness is illustrated in Figure 10.

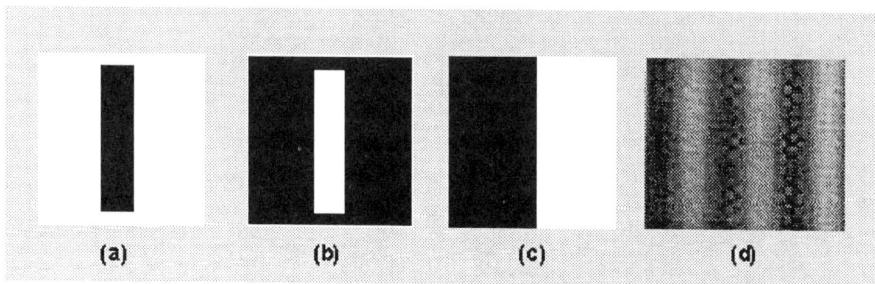

FIGURE 9.

Examples of simple stimuli used in investigations of the sensitivity of neurons: black bar (a), slit of light (b), visual edge (c), and sinusoid grating (d).

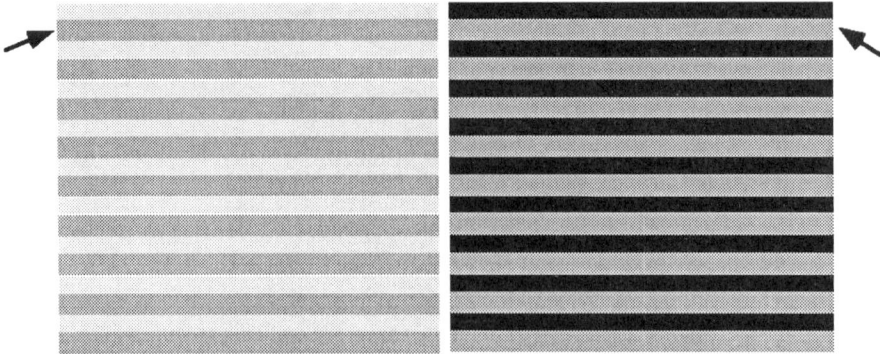

FIGURE 10.
Dark and light gray stripes are superimposed on the same gray background. However, the background associated with the lighter-gray stripes appears darker than the background of the dark stripes, although the intensity of reflected light is constant.

Color (or more correctly *hue*) is another example of a perceived stimulus attribute that is related to differences in the absorption characteristics of reflecting surfaces. In natural viewing conditions light sources irradiate light across a wide range of wavelengths, and surfaces differ in their ability to reflect different wavelengths. The physical correlate of hue is wavelength. The wavelength composition of light reflected from a surface depends upon the wavelength composition of the light reaching that surface and upon its reflective properties. The wavelength of light reflected from a surface does not bear a fixed relation with the perceived hue of that area. Usually, short light wavelengths are perceived as blue, middle wavelengths as green, and long wavelengths as red. However, perceived hue may remain constant despite large variations in the wavelength composition of the reflected light (produced by corresponding changes in the wavelength of the light incident upon the surface), a phenomenon known as *color constancy*. For instance, a red surface will be perceived as such, even when the light reflected from it contains a greater proportion of short wavelengths than long wavelengths (Land, 1974). The visual context in which the surface in question is presented plays a crucial role in this phenomenon: if the same area were presented in isolation rather than as part of a multicolored display its hue would be perceived as green. Thus, across different lighting conditions, the visual system must be capable of "translating" contradictory wavelength cues in order to assign the "correct" color to a given surface. As we shall see in the following sections, most neurons in the visual system are only sensitive to the wavelength composition of the light reflected from various surfaces. However, neurons in at least one visual cortical area appear to be sensitive to the hue of a particular surface as perceived by human observers, rather than the wavelength composition of the light the surface reflects per se (Zeki, 1983a, b).

In addition, local brightness contrasts that compose a typical visual scene can undergo changes in time. In experimental settings, the simplest form of temporal change is that of a spot of light that is switched on and off repeatedly at regular intervals. The rate

of such changes in brightness is described by *temporal frequency*, which is operationally defined as the number of complete light/dark cycles per second or Herz.

Another aspect of vision entails the capacity to estimate the distance of visual objects
from the observer and to approximate their relative position in depth. The perception of
depth can be achieved by sensory cues that require binocular viewing, and also by a variety of monocular sensory cues (like motion parallax, shading, interposition, linear perspective, etc.). The notion that binocular vision plays an important role in the perception
of depth has existed for more than one and a half centuries (Wheatstone, 1838). This
feature of binocular vision is due to the fact that the eyes are located at a distance from
one another. As illustrated in Figure 11, viewing an object with both eyes creates images
on the two retinas that are slightly displaced with respect to the center of each retina.
This displacement creates a variable amount of (binocular) horizontal *retinal disparity*.

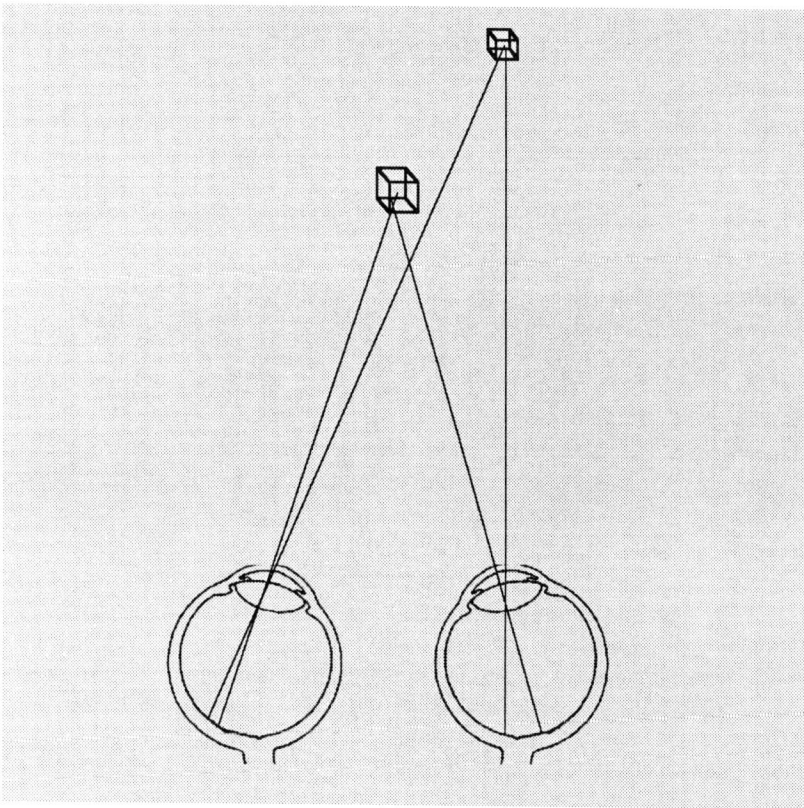

FIGURE 11.

Relative binocular disparity of two objects placed at different distances in front of the observer.
The distance between the images of the two objects on the retina of the right eye is smaller than
the corresponding distance on the retinal of the left eye. The interocular difference in the separation of the two images is equivalent to the relative disparity of the two objects, which is translated
in a difference in the perceived relative depth of the two objects.

A related phenomenon is *binocular fusion*, the process whereby the images produced on the two retinas by a single object are integrated and trigger a unified percept. This capacity requires that the visual system can establish the correspondence between retinal points in the two eyes that receive the same portion of the projected image of a given object. Fusion and stereopsis represent two complementary capacities of the visual system. Fusion is optimally achieved when the retinal disparity is either very small (less than approximately 20 min. of arc for foveal vision) or zero (i.e., with a single visual target). On the other hand, the presence of small disparities between the two retinal images is a prerequisite for stereopsis.

The capacity of neurons in the visual system to convey information regarding depth cues has been studied with a variety of experimental paradigms. Initially, stimuli that contained both binocular and monocular cues were used, like the one shown in the right-hand portion of Figure 12. In this case the visual array consists of two gratings, each projected to one eye. The two gratings are identical, but one is shifted laterally with respect to the other, by a few minutes of arc. Thus, each vertical line, by falling on non-corresponding parts in each retina, generates an interocularly disparate image. However, even when the two gratings are viewed monocularly, non-stereoscopic depth cues are detectable, such as motion parallax produced by lateral head movements. In situations like this, depth perception is possible even by an individual who cannot process binocular disparity cues.

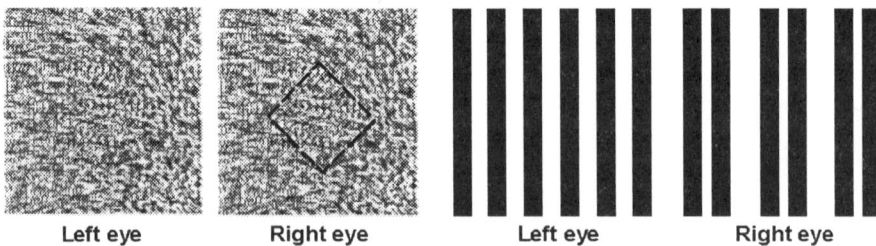

Left eye   **Right eye**   **Left eye**   **Right eye**

FIGURE 12.
(Left) Computer generated random-dot stereogram. Under natural viewing conditions, either with one or with both eyes, it appears as a random array of colored dots. When viewed with chromatic filters each eye receives the same matrix, but each element of the matrix stimulates a slightly different retinal location in each eye. This retinal disparity, which is restricted to elements within a rhomboid area in the center of the array, produces the perception of a diamond-shaped surface that "floats" over the background (Adapted from Julesz, 1978). (Right) Example of a line stereogram that contains both monocular and binocular depth cues. The left eye is stimulated with a regular grating pattern. In the stimulus directed to the right eye alternate bars have been shifted laterally to create binocular horizontal disparity. When viewed with polarizing glasses, the second, fourth, and sixth bars on one side appeared to stand out of the screen. (Reprinted from Birch et al., *Investigative Ophthalmology and Visual Science, 26*:366-370, 1985 with permission from the Association for Research in Vision and Ophthalmology).

To overcome these limitations, visual scientists constructed stimuli that contain depth cues detectable exclusively under binocular viewing conditions. These computer-generated stimuli, known as *random dot stereograms*, consist of two identical matrices of randomly distributed dots, a red-black and a green-black matrix that are overlaid on each other (see left-hand portion of Figure 12). In order to isolate the images produced by each matrix and project each image to one eye only (in other words introduce *dichoptic vision*), the subject wears chromatic filters so that one eye sees only the red-black matrix while the other eye receives only the green-black matrix. A subset of the dots in each matrix, arranged in such a way as to form a recognizable shape (e.g., a square), is placed with a slight horizontal offset. The two embedded patterns are shifted laterally, relative to each other, so that corresponding points in each shape are projected on slightly disparate (non-corresponding) retinal loci. Observers who possess stereoscopic vision report seeing a form that "pops out" in front of the display. In a further improvement of this technique (Julesz, 1978; 1986), the stimulus elements are generated on a computer monitor and refreshed at a very high rate (*dynamic random-dot stereograms*). In this way any monocular cues are eliminated. The perception of stereoscopic forms in these stimuli is known as *global stereopsis*.

It should be emphasized that the neural operations that lead to visual perception do not necessarily take into account the stimulus parameters described thus far. Therefore, finding neurons that respond to variations of a particular *sensory cue*, such as a brightness contrast, does not necessarily imply that the activity of these cells provides the neural substrate for perceiving a stimulus attribute. Contrast-sensitive neurons may be found in many functionally distinct areas. Each area may utilize brightness contrast cues in order to extract a different stimulus attribute such as shape and texture. Moreover, a particular stimulus attribute can be determined by more than one visual cue. In our example, retinal disparity can convey information regarding visual shape, in addition to contrast. Each of these cues may be analyzed by a different population of neurons, possibly in different regions of the visual system. Nonetheless, the sensory cues described above serve as useful abstractions that enable researchers to use a universally accepted terminology when referring to experimental stimuli and conditions. The goal of *visual psychophysics* is to study how the properties of the visual world are related to subjective experience. Traditionally, psychophysics is concerned with establishing the limits of conscious sensation, known as psychophysical *thresholds*. In the simplest paradigm, the experimenter varies one parameter in a spot of light (such as intensity, wavelength, duration, or size) while keeping the remaining parameters constant, and asks the observer to press a button when he/she first detects the stimulus. Although these experiments can provide some information regarding the sensitivity of the visual system to simple visual stimuli, they do not provide any information as to how the visual system encodes the spatial distribution of light variations. As previously mentioned these variations, or local brightness contrasts, are the elementary components of natural visual scenes. The sensitivity of the visual system to brightness contrasts is described by the generic term: *spatial resolution*. There are several measures of spatial resolution, two of which are of some relevance to our discussion: *visual acuity* and *contrast sensitivity*. The former is a measure of the ability of the visual system to resolve closely spaced bars that make up a high contrast contour, known as a grating.

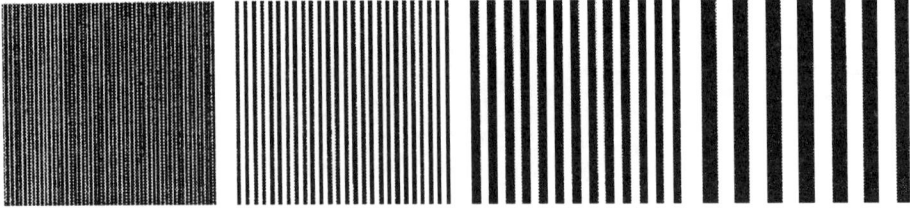

FIGURE 13.
Four sinusoid gratings at progressively lower spatial frequencies (from left to right).

The visual acuity threshold reflects the maximum spatial frequency of a visual grating (or alternatively, the minimum separation between adjacent bars) that an observer can resolve (see Figure 13). The visual acuity threshold is measured in cycles per degree of visual angle cycles/degree). For the average observer the upper limit of spatial resolution under optimal contrast conditions is 30 cycles/degree for central vision. Visual acuity drops dramatically at progressively larger eccentricities.

However, under natural viewing conditions the contrast between adjacent parts of a visual scene is not always optimal. Natural scenes contain luminance transitions that vary in both contrast and spatial distribution. To address this problem, psychophysicists came up with an alternative measure, the contrast sensitivity function, which is a plot of acuity thresholds over a wide range of stimulus contrasts. An example of the contrast sensitivity function that is typical of adult observers with normal or corrected-to-normal vision is presented in Figure 14. Visual acuity and contrast sensitivity both depend heavily upon two factors: the average luminance of the visual stimulus and the angular distance of the stimulus from the center of the array of specialized photoreceptors on the back of the eye (i.e., retinal eccentricity).

In humans, estimates of spatial resolution can be obtained in several ways that fall into two broad categories: psychophysical methods and non-invasive electrophysiological methods. In the former, the observer indicates that he/she has detected the stimulus by making a behavioral response (button press). Some of these methods have been adapted for use with infants and young children (Teller, 1979) and animals (Mitchell et al., 1976). In these procedures, the tester monitors one or more behaviors as indicators that the infant has detected the patterned stimulus. Electrophysiological methods rely upon the measurement of certain parameters of the visual evoked response (VER), which reflects a change in the electrical potential measured from the scalp in response to a complete phase reversal of a bar grating. In this test, dark areas become white and white areas become black at a rate of 12-20 reversals per second.

The evoked responses display a characteristic sinusoid morphology with each contrast reversal being associated with a single peak or trough in the VER waveform, as shown in Figure 15. This method produces contrast sensitivity functions that are very similar to those obtained with traditional behavioral methods, and can provide an objective method of determining the spatial resolution of the visual system during the early stages of development.

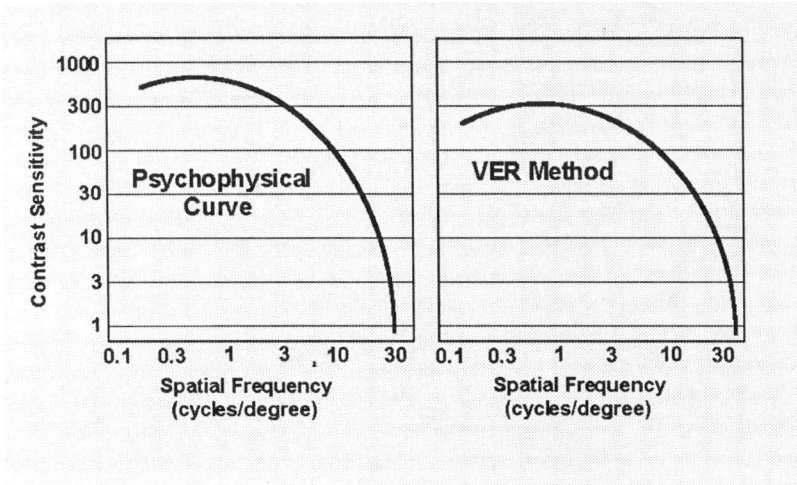

FIGURE 14.
Average contrast sensitivity function obtained from seven adult observers using two methods. The psychophysical, on the left, was obtained by determining the absolute contrast threshold for detecting a sinusoid grating of a particular spatial frequency. In this test, observers indicate behaviorally when they detect the stimulus. The examiner then repeats the experiment for several different spatial frequencies. In the graph, contrast sensitivity is the reciprocal of the threshold value obtained for each grating. The intersection of the curve with the x axis corresponds to the visual acuity threshold. (Reprinted from *Vision Research, 30*, Norcia et al., pp. 1475-1486, Copyright 1990, with permission from Elsevier Science).

An estimate of the capacity of the visual system to resolve light stimuli presented in rapid succession is given by the *flicker fusion threshold*: which corresponds to the maximum temporal frequency at which rapidly flashing stimuli are perceived as separate events rather than a continuous light. Finally, the sensitivity of the visual system to retinal disparity cues is known as *stereoacuity*. The stereoacuity threshold indicates the minimum amount of retinal disparity produced by the images of two objects placed in front of the observer, that is necessary to perceive them as located at different relative depths. Adult observers with normal stereoscopic vision have stereoacuity thresholds in the order of a few seconds of arc.

As mentioned above, the immediate goal of psychophysics is to determine the range of sensory capabilities in a given species. Thus, psychophysical data provide measures of the sensitivity limits of the visual system.

The ultimate goal of this line of research is, however, to identify the operations or mechanisms that are responsible for the observed sensory capabilities. To accomplish this goal, psychophysicists construct models that specify the characteristics of the neural systems believed to be involved in a given sensory process.

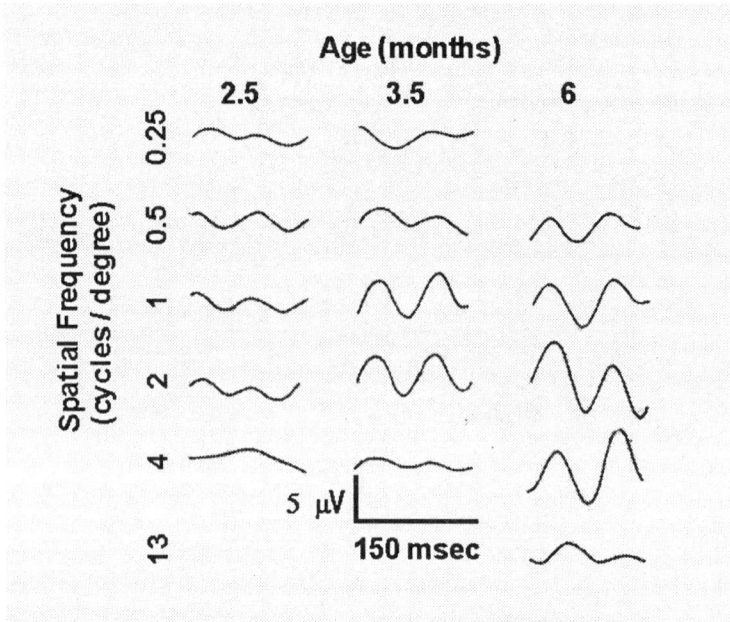

FIGURE 15.
Visual evoked responses obtained from the same infant at 2.5, 3.5, and 6 months of age. The
VERs were elicited by a sinusoid grating that was changing phase 16 times per second (phase reversal frequency = 8 Hz). The infant sat on her mother's lap and simply looked at the projection
screen while recordings were made from an electrode placed near the inion (over the occipital
lobe). Electrical response epochs to 50 reversal cycles were obtained and averaged together to
improve the signal-to-noise ratio, and the averaged waveforms were subsequently filtered in order
to emphasize frequencies around the stimulus reversal frequency. This procedure was repeated
with gratings of increasingly higher spatial phase at different brightness contrast levels. Notice
that, for a given age, the amplitude of the response (i.e., the voltage difference recorded between
the inion and a "neutral" electrode placed on the top of the infant's head) is maximal at a particular spatial frequency. This spatial frequency increases with age, reflecting a corresponding improvement in spatial resolution. (Reprinted from *Brain Research, 141*, Pirchio et al., Infant contrast sensitivity evaluated by evoked potentials, pp. 179-184, Copyright 1978, with permission
from Elsevier Science).

Such models permit scientists to predict the output of these hypothetical systems in
terms of behavioral performance parameters. Subsequently, they set up experiments that
involve the manipulation of a set of stimulus parameters and use the psychophysical data
derived from these experiments to test the validity of their predictions. However, the
number of conclusions that have been drawn on the basis of psychophysical data is disproportionately small relative to the volume of research that has been conducted since
the founding of the field more than a century ago. The most severe limitation of psychophysical research based upon behavioral measures is that, in principle, it operates
without knowledge of any of the properties of the neural system that it strives to reveal.

With the development and widespread use of single-cell recording techniques, vision scientists have been able to examine the sensitivity of individual neurons in the visual pathway. More than three decades have passed since the first demonstration of stimulus-specific response properties of individual neurons in the primary visual cortex (Hubel & Wiesel, 1962). In extracellular recordings from the visual cortex of the cat, Hubel and Wiesel described cells with receptive field properties that progressively increased in complexity from one class of cells to another. In order to account for these findings, they proposed a model based on hierarchical connections between cells from different classes (i.e., simple, complex, and hypercomplex), which, after undergoing several modifications, survives to this day. Such models seek to explain how the visual array is analyzed into elementary features, which can then be integrated to provide a neural representation of more complex stimulus attributes.

At approximately the same time that Hubel and Wiesel made their discoveries, data from a number of studies strongly suggested that visual input is transmitted through parallel and relatively independent physiological channels in the retinocortical pathway. At the level of the retina, this property was acknowledged very early with the first reports of physiologically distinct groups of ganglion cells in the cat retina (Enroth-Cugell & Robson, 1966). Other studies revealed the presence of a similar functional differentiation in the visual thalamic nuclei (i.e., the dorsal lateral geniculate nucleus; De Valois et al., 1966). Moreover, there was evidence that these distinct cell types project to different targets in the primary visual cortex (Hubel & Wiesel, 1972). A few years later, cortical neurons with distinct response selectivities were discovered in primate visual association areas (Zeki, 1973, 1974).

In order to take into account these findings, current models of visual system organization emphasize the existence of anatomically and functionally distinct subsystems, each principally involved in the analysis of different aspects of visual input. At the level of the association cortex, one subsystem is thought to be primarily involved in the perception of spatial relations and visual motion, and in the guidance of tracking eye movements. A second pathway appears to be specialized for the operations that lead to object recognition. Results from animal lesion studies corroborate this distinction (Maunsell & Newsome, 1987; Merigan & Maunsell, 1993). In humans, functional dissociations have been noted for different aspects of visual perception, such as for processes leading to object recognition and to the perception of spatial relations and visual movement (Zeki et al., 1991; Zihl et al., 1991).

The goal of the second part of the book is to provide an empirical framework for the neurophysiological mechanisms that presumably mediate basic perceptual abilities. The first section contains a description of the geniculocortical pathways followed by a discussion of the physiological properties of the most prominent cell classes encountered in these structures and their organization in the primary visual cortex. In the next two sections the focus of the discussion will be on visual association areas that have been identified in the primate occipital, parietal, and temporal neocortex. Also included is a brief discussion of the specific impairments underlying disorders of object and face recognition in humans. The last section is devoted to an appraisal of current views regarding the relation between physiologically distinct retinocortical pathways and cortical subsystems, and the role that visual association areas play in the analysis of visual cues.

NOTES

[1] The simplest form of a brightness transition is a *visual edge*. Spatial variations in light intensity can also be described in terms of spatial frequency. According to this notion any visual pattern can be decomposed into a finite set of sinusoid frequency components through Fourier analysis procedures. Each of these components can be represented by a periodic pattern of alternating stripes, characterized by a smooth transition between bright and dark regions. Such a pattern can be fully described by its spatial frequency, i.e., the number of bright-dark cycles per degree of visual angle.

Chapter 4

Retinocortical Pathways

The role of the eye as an optical apparatus is to transmit, with minimal distortion, images of the external world to the specialized photoreceptors that are distributed along the posterior part of the eye bulb (i.e., the retina). A number of structures, located in the anterior part of the eye, contribute to this task. These include the cornea, iris, cilliary muscle, and lens (see Figure 16). The latter is a biconvex structure capable of changing shape through the action of the cilliary muscle. The variable shape of the lens ensures that objects located at different distances from the observer form sharp images on the retina (i.e., images with minimal blur or defocus). The light passes through the cornea, the lens and the vitreous humor to reach the retina. The latter, as shown in Figure 16, contains a layer of photoreceptor cells, and several other layers that contain first- and second-order afferent neurons (the bipolar and retinal ganglion cells, respectively), and two main types of interneurons: horizontal and amacrine cells. It is important to remember that the photoreceptor cell layer forms the outer border of the retina. Thus, the incoming light must travel through all the other retinal layers before it reaches the photoreceptors. In a small circular region of the retina, the fovea,[1] the internal cell layers are displaced laterally, allowing light to reach the photoreceptors with minimal distortion. The fovea, and especially its central region, the foveola, contains the highest density of photoreceptors than any other

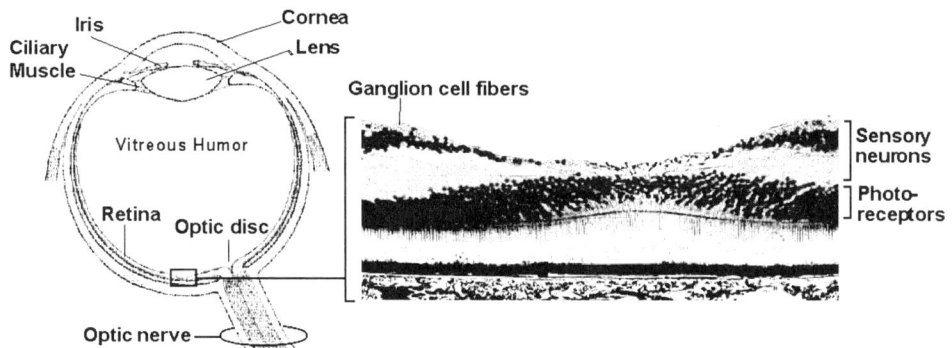

FIGURE 16.
(Left) Schematic horizontal cross-section through the left eye (top view). (Right) Photomicrograph of the human retina (cross section through the foveal region). (Reprinted from *Progress in Neurobiology*, 54, Provis et al., Ontogeny of the Primate Fovea: A Central Issue in Retinal Development, 549-581, Copyright 1998 with permission from Elsevier Science).

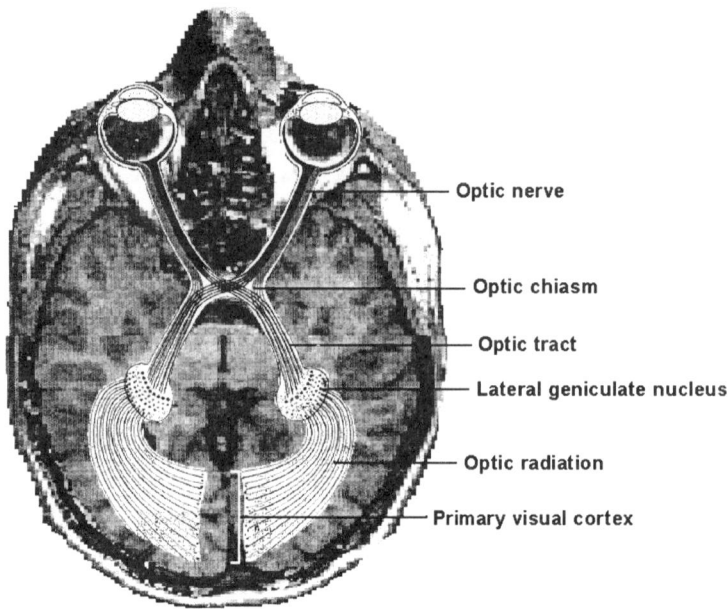

FIGURE 17.
Horizontal sections through the human brain at two different levels combined to show the entire extent of the pathways from the retina to the primary visual cortex.

region of the retina. The photoreceptors are specialized organs capable of responding to light particles (photons) with a change in their membrane potential. They synapse with bipolar cells which, in turn, form contacts with retinal ganglion cells. The axons of the latter exit the eye at the optic disc and, after acquiring a myelin sheath, form the optic nerve. The optic nerves (one from each eye) decussate partially in the optic chiasm, which is located at the base of the brain, just rostral to the hypothalamus (see Figure 17). After leaving the optic chiasm, the ganglion cell fibers form two optic tracts. Each optic tract contains fibers from the temporal half of the ipsilateral retina and from the nasal half of the retina in the contralateral eye. The majority of these fibers project to the dorsal lateral geniculate nucleus (dLGN) on each side.

The dLGN contains six well-defined cellular laminae (see Figure 18). Retinal inputs from each eye remain segregated in the primate dLGN: fibers from the ipsilateral eye project to laminae 2, 3, and 5, whereas fibers from the contralateral eye terminate in laminae 1, 4, and 6. The projections to the dLGN are *retinotopically* organized within each lamina: adjacent points in one hemiretina project to adjacent points in the nucleus. Moreover, the retinotopic maps found in each layer are in precise register with each other. An electrode inserted into the nucleus perpendicular to the plane of the laminae encounters cells with receptive fields in corresponding points of the visual field. This arrangement is important for sustaining binocular vision, namely, that images of the same object, sampled simultaneously by both eyes, produce a single percept (i.e., they

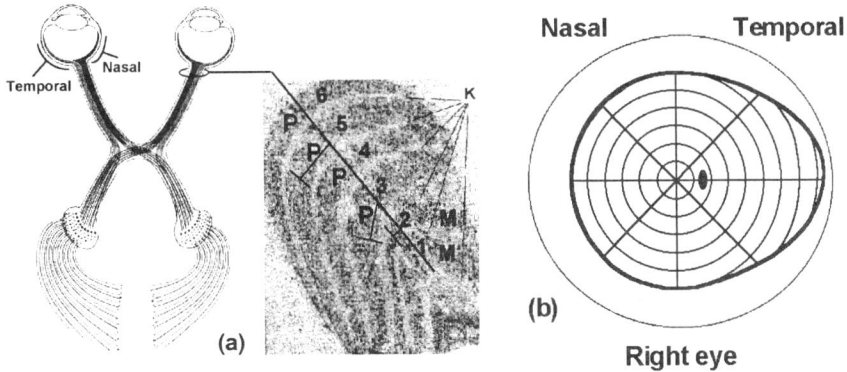

FIGURE 18.
(a) Cross-section through the right dorsal lateral geniculate nucleus. The Nissl technique reveals the arrangement of cell bodies in six layers. Layers 1 and 2 (M or magnocellular layers) receive input from one subpopulation of retinal ganglion cells, whereas layers 4-6 (P or parvocellular layers) receive input from a different subpopulation of retinal ganglion cells. Recently, 6 additional cell layers have been found in the dLGN that contain cells with very small bodies and are referred to as K (or koniocellular) layers. These cells contain three proteins that are found exclusively in koniocellular layers. (Reprinted from *Trends in Neuroscience, 22,* Hendry & Calkins, Neural chemistry and functional organization in the primate visual system, pp. 344-349, Copyright 1998, with permission from Elsevier Science). (b) The spatial layout of the visual field of the right eye is indicated by the solid line. The dark area indicates the location of the blind spot, that is, a small region on the retina where optic nerve fibers converge before they exit the eye. Notice that the visually accessible area extends more laterally (i.e., into the temporal hemifield) than it does medially (i.e., into the nasal hemifield). Only the area enclosed by the solid line is represented in layers 2, 3 and 5 in the dLGN.

are perceived as fused, see Figure 19 and Chapter 5 for more details). More details about the mechanisms of binocular vision are given in Chapter 5. The dLGN relays retinal input to the superior colliculus, as well. Projections to the superior colliculus originate from a relatively small population of neurons located between the main geniculate laminae (koniocellular layers). In addition, the superior colliculus receives direct input from retinal ganglion cells. These projections play an important role in the control of rapid eye movements (saccades).

The afferents from each dLGN form the optic radiations that arch around the lateral ventricle to reach the primary visual cortex (Brodmann's area 17 in humans or area V1 in primates, see Figure 20). A cross-section of the visual cortex reveals (shown on the right-hand portion of Figure 20) six main layers that show distinct cytoarchitectonic characteristics. They also show distinct patterns of afferent and efferent projections. A number of fibers terminate in the adjacent visual association areas (Brodmann's areas 18 and 19 which correspond to areas V2 and V3 in primates, respectively). Area V1, which lies on the banks of the calcarine fissure on the medial surface of the occipital lobe in primates,

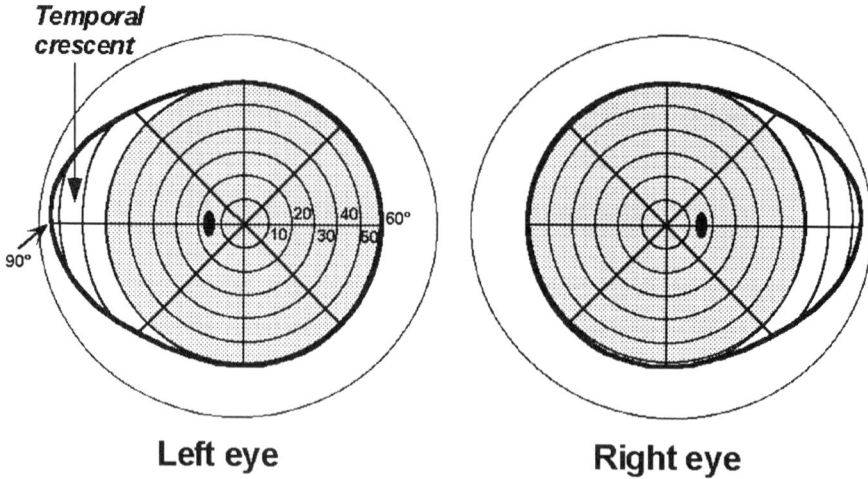

Left eye   **Right eye**

FIGURE 19.

Layout of visual fields for each eye. The visual field extends up to approximately 60 degrees of eccentricity in the nasal direction and up to about 90 degrees in the temporal hemifields. The shaded area indicates the binocular portion of the visual field.

is also known as the striate cortex. It owes its name to a thin layer of myelinated fibers -- visible in a cross section of the cortex -- traversing layer IV.

This layer receives the geniculocortical afferent fibers. Projections into layer IV maintain the retinotopic organization observed in dLGN.

The visual cortex in each hemisphere contains a representation of only the contralateral visual field. Signals from the lower half of the retina terminate in the upper bank of the calcarine fissure, and signals form the upper half project to the lower bank. Fibers that carry input from the fovea terminate in the posterior part of area V1. The foveal representation occupies areas in the dLGN and the primary visual cortex that are disproportionately large with respect to the actual size of the fovea. This phenomenon, which is known as *magnification*, reflects in part the higher density of photoreceptors in the central fovea than in the periphery of the retina. In addition, there is no convergence in the synapses between photoreceptors and retinal ganglion cells in this region. Cortical magnification further enhances this tendency and contributes to the fact that spatial resolution is higher, by many orders of magnitude, for images projected on the fovea than for peripheral vision. Another correlate of cortical magnification is the extensive *divergence* of afferents in each relay station along the visual pathway. In the cat, each retinal ganglion cell synapses with approximately four cells in the dLGN, and each geniculate cell with approximately 25 cortical cells. In primates there is minimal divergence between retinal ganglion cells and dLGN neurons, but this is compensated later on because each dLGN neuron forms synapses with an average of 56 cells in area V1 (Peters et al., 1994). Another phenomenon that affects the distribution of retinal input to the cortex is *convergence*. It is estimated that each neuron in layer IVC in area V1 in primates receives input from an average of eight dLGN cells. Therefore, it is unlikely that the firing pattern of

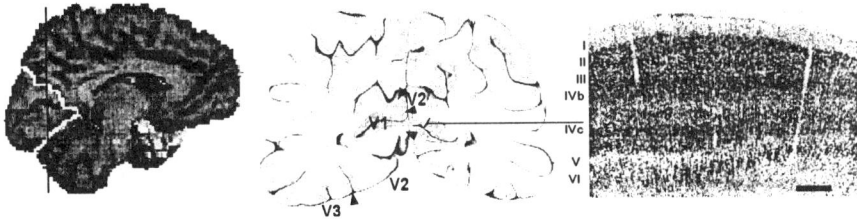

FIGURE 20.
(Left) Midsagittal view indicating the location of the calcarine fissure (dark trace) in the human brain. (Middle) Coronal section through the occipital lobe at the level indicated by the vertical line on the left-hand image. (Right) Nissl-stained section through area V1, revealing the distribution of cell bodies in six main layers.

any neuron in area V1 that receives geniculate input is determined solely by the input of any single geniculocortical afferent.

A gross index of the relative importance of vision compared to the other modalities in primates is the fact that visual areas occupy a large portion of the neocortical surface (up to 40% of the total brain surface in some species[2]). By combining anatomical, electrophysiological, and experimental lesion data, researchers have identified over 30 separate neocortical visual areas in a number of primate species (Van Essen, et al., 1991). Figure 21 presents the areas discussed in this book. These areas receive input relayed predominantly via the primary visual cortex and contain neurons that respond to visual stimulation. Some of these areas also receive non-visual input, such as input from other afferent modalities and input regarding the static and kinetic state of the eyes. Each of these areas shows some form of retinotopic organization. As a rule, the mapping of the visual field is very precise and orderly in the most caudally located areas and becomes progressively more diffuse as one moves rostrally. This phenomenon is related to the fact that the size of the receptive field of neurons increases as one moves from the occipital pole to areas located more rostrally. Naturally, it becomes increasingly more difficult to define the topography of the retinotopic map when the receptive field of most cells is large enough to include a major portion of the visual field.

After reviewing a large body of research regarding connectivity patterns, cytoarchitectonic characteristics, and retinotopic organization, Felleman and Van Essen (1991) found strong evidence in favor of the traditional notion that the visual areas are organized in a hierarchical manner. Nonetheless, from the 305 identified pathways interconnecting visual areas, the large majority are bi-directional connections. In addition, there are numerous "lateral" connections, i.e., connections between areas which, on the basis of a combination of the three criteria mentioned previously, can be assigned to the same level in the hierarchy. In this chapter we will concentrate on the following visual areas: V1, V2, V4, Middle Temporal area (V5), Middle Superior Temporal area (MST), and Inferior Temporal area (IT). With the exception of the first two, each of the following areas has been found to contain a number of subregions. For instance, area MST has two subregions (the dorsomedial [MSTd] and the dorsolateral MST [MSTl]). Six subregions have

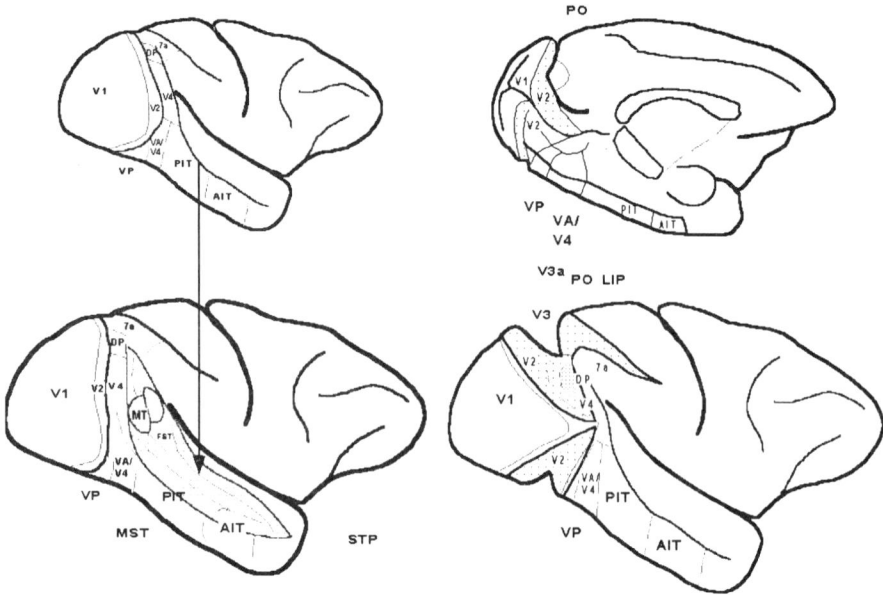

FIGURE 21.

Top row: lateral (left) and mesial (right) view of the right hemisphere of the macaque brain indicating the location of visual areas. Bottom row: The location of areas buried within sulci. Abbreviations: MT, Middle Temporal area; MSTd, dorsal Middle Superior temporal area; PIT, Posterior Inferior Temporal area; AIT, Anterior Inferior Temporal area; STP, Superior Temporal Polysensory area; VP, Ventral Posterior area; PO, Parieto-occipital area; DP, Dorsal Prelunate area, FST: Fundal Superior Temporal area, LIP: Lateral Intraparietal area, VOT: Ventral Occipito-temporal area. (Reprinted with permission from the *Annual Review of Neuroscience, Volume* 10, Copyright 1987 by Annual Reviews www.AnnualReviews.org).

been identified in area IT: anterior, central, and posterior regions in the dorsal and the ventral aspect of IT (AITd, CITd, PITd, AITv, CITv, and PITv, respectively).

It should be noted that in primates a single cytoarchitectonic area may contain more than one functionally and anatomically distinct area. For instance, areas V2, V3, V3a, and V4 belong histologically to Brodmann's area 18 (Zeki, 1978).

The retina: Anatomy and physiology

Phototransduction takes place in the retina which forms the inner layer of the eye. There are five main classes of neurons in the mammalian retina. Photoreceptors, bipolar, and retinal ganglion cells form the main transduction pathways, while important regulating circuits are formed by two main classes of interneurons: horizontal and amacrine cells. Several subtypes have been identified within each of these populations, including at least ten bipolar cell subtypes, 20-25 retinal ganglion cell subtypes, and up to 40 different types of amacrine cells. With the exception of horizontal cells, only a fraction of these

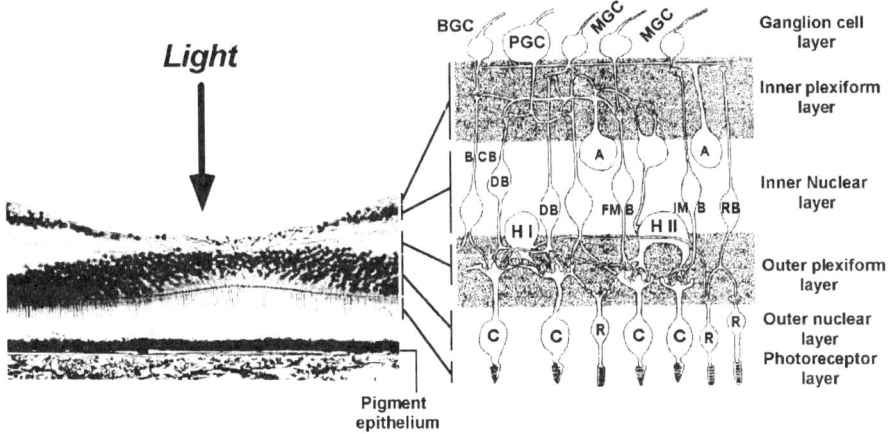

FIGURE 22.
(Right) Cross section through the adult human fovea (from Fig. 16). Notice the thinning of the ganglion cell and inner plexiform layers, that is especially pronounced in the foveola, the central region of the fovea. Cone density reaches its maximum value in this region. The characteristic curvature displayed by the inner surface of the retina in the foveal region is known as the foveal slope. (Left) simplified wiring diagram of the retina. Abbreviations; C: cone, R: rod, HI and HII: Type I and Type II horizontal cells, respectively, DB: diffuse bipolar cell, IMB, FMB: invaginating and flat midget bipolar cells, respectively, BCB: blue cone bipolar cell, A: amacrine cell, MCG, BGC, and PGC: midget, bistratified, and parasol ganglion cells, respectively. (Reprinted with permission from Hickey & Peduzzi, 1987, *Handbook of Infant Perception, I.* Copyright 1987 by Academic Press).

cell subtypes have been adequately studied and characterized. These will be discussed in later sections.

These cells are arranged in several layers shown in Figure 22. The outermost layer contains the outer segments of the photoreceptors (photoreceptor layer), the cell bodies of which are located in the outer nuclear layer. Cell processes that connect photoreceptors, bipolar, and horizontal cells form the outer plexiform layer. Moving toward the center of the eye one finds the inner nuclear layer which contains the bodies of bipolar, horizontal and amacrine cells. The synapsing processes of bipolar, amacrine, and retinal ganglion cells are, in turn, arranged along the inner plexiform layer. The innermost retinal layer is occupied by the axons of retinal ganglion cells that extend toward the back of the eye, where they exit the globe at the optic disc, forming the optic nerve. The retina is transversed by a special type of glial cell, the Müller cell, which fills most of the extracellular space in the retina. A simplified wiring diagram of the retina is presented in Figure 22. Photoreceptors make contacts with second-order sensory neurons (bipolar cells) and interneurons (horizontal cells) in the outer plexiform layer. Bipolar cells, in turn, transmit photoreceptor signals to third-order sensory neurons (retinal ganglion cells) and to a different type of interneuron located in the inner plexiform layer (amacrine cells). Neuronal signals are transmitted to the brain via the optic nerve which consists exclusively of the axons of retinal ganglion cells.

The outer segments of photoreceptors are positioned immediately against a thin layer of epithelial cells known as the pigment epithelium. The latter serves a number of supporting functions for the neural retina which include: (1) the establishment of a blood-retina barrier controlling the flow of large molecules and ions in and out of the retina, (2) disposal of spent outer segment tips through phagocytocis, and (3) absorption of light particles that have not entered into the apertures of photoreceptor cells by the light absorbing pigment, melanin. Light must travel through each of the neural retinal layers in order to reach the photoreceptors, the outer segments of which are pointing away from the lens. To ensure that light reaches the foveola cones with minimal distortion, the foveal floor shows a depression as neural elements normally located in the inner retinal layers are displaced radially. Synaptic contacts formed by foveola cones are also displaced radially and carried out by vastly elongated cone axons.

Phototransduction

Phototransduction refers to the processes through which light falling on the retina is converted into electrical signals. This process sets the limits of crucial properties of vision like, *absolute sensitivity, temporal resolution,* and *spectral or wavelength sensitivity.* A characteristic of all photoreceptors is that their membrane remains in a depolarized state in the dark, releasing glutamate constantly from their synaptic endings. In the dark, the membrane potential is generated primarily by inward flow of cations (mostly Na^+ and Ca^{2+} forming the "dark current") through ion channels located in the outer segment. These channels are gated by cyclic guanosine monophosphate (cGMP): they are kept open at high intracellular concentrations and close at low concentrations of cGMP. When photoreceptors are stimulated by light, these channels close leading to membrane hyperpolarization and a reduction in transmitter release. The sequence of phototransduction events in rods can be briefly described as follows (phototransduction in cones is based on a similar mechanism): photons captured in the outer segment trigger the conversion of rhodopsin to an active form (Meta II rhodopsin). Metarhodopsin, in turn, initiates a series of chemical events that lead to the hydrolysis (and deactivation) of cGMP by the enzyme phosphodiesterase. Reduced intracellular level of cGMP closes cationic channels and, as the inward flow of positively charged ions decreases, membrane potential moves to more negative values (hyperpolarization). Photoreceptors become resensitized as the photoexcited form of rhodopsin is converted back into its photosensitive form. In addition, the membrane returns to a depolarized state as cGMP-gated cation channels reopen. This is caused by: a) an enzymatic reactivation of cGMP that is permitted by reduced levels of intracellular Ca^{2+} leading to increased intracellular levels of cGMP and, b) increased sensitivity of these channels to cGMP that is also associated with reduced Ca^{2+} levels.

Photopigments

Photoreceptors differ in their sensitivity to wavelength by virtue of the fact that they contain only one of a set of four visual pigments. There are three types of cone pigment: blue, green, and red, each maximally absorbing light at approximately 440, 530, and 560 nm. There is one type of rod pigment, rhodopsin. Cone photoreceptors that contain blue pigment are often referred to as S cones to indicate the fact that they show maximal sensitivity to short wavelengths. Accordingly, green- and red-pigment containing cones are called M for medium wavelength and L for long wavelength. The spectral sensitivity

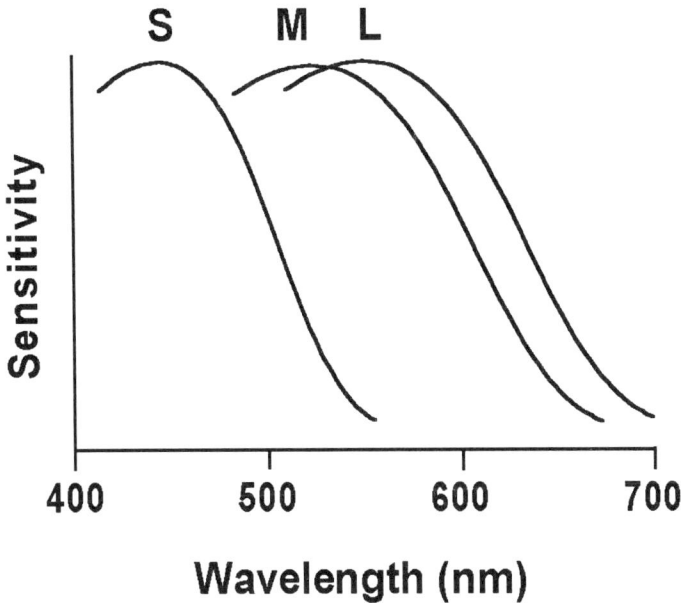

FIGURE 23.
Wavelength sensitivities of macaque cone photoreceptors determined electrophysiologically.
(Based on data from Baylor, 1992).

of different cone types matches closely the spectral absorption properties of the pho-
topigments they contain, as illustrated in Figure 23.

Visual pigments consist of a long polypeptide chain, opsin, that traverses the photoreceptor membrane seven times. Opsin is bound to a photosensitive chromophore, *retinal*,
which when stimulated by light is converted from an *11-cis* to an *all-trans* form. This
conversion is known as photoisomerization.

The genes that encode human visual pigments have been isolated on chromosomes 3
and 7, for rhodopsin and the blue pigment, and on the X chromosome, for red and green
pigments. Individual variations in sensitivity to color can be, at least partly, explained by
small, genetically determined, differences in the wavelength absorbance properties of
certain pigments (Winderickx et al., 1992). Moreover, deficits in color vision can be explained by the lack of one or more of these genes or by the presence of atypical (hybrid)
genes (for a review, see Nathans et al., 1992). The most common color vision defects are
X-linked anomalies of red/green vision, estimated to occur in approximately 8% of Caucasian males and only 1% of Caucasian females. These estimates are lower in other
populations. In some cases either the red or the green mechanism is entirely missing (*di-
chromacy*), whereas in other cases spectral sensitivity is somewhat shifted without complete loss of the ability to detect the color in question (*anomalous trichromacy*). Other,
much rarer conditions, include *rod monochromacy* (an autosomal recessive trait in which
vision is mediated exclusively by rods), *blue cone monochromacy* (an X-linked anomaly

in which red and green cones do not function), and *congenital tritanopia* (an autosomal trait characterized by serious impairment of the blue cone mechanism). In blue cone monochromacy, rods mediate vision at scotopic levels, while at higher light levels vision is mediated primarily by blue cones. Interestingly, at intermediate light levels these patients retain rudimentary hue discrimination ability. They show, however, a serious reduction in visual acuity which is attributed to the fact that the central fovea area contains only the defective red and green cones. Tritanopia, on the other hand, is usually secondary to dominant juvenile optic atrophy.

Classifications of retinal cells
Rods, with their characteristically thin tubular inner and outer segments (see Figure 24), constitute the majority of photoreceptor cells in the human fovea -- approximately 80%. Cones derive their name from the conical shape of their inner segment. There is considerable variability in both the shape and size of cones with eccentricity, for instance foveal cones are thinner than peripheral cones by an order of magnitude. While rod photopigment discs in the outer segment are separated from the extracellular space by the cell membrane, photopigment discs in cones are not. Maximum rod density in the human fovea occurs between $20°$ and $30°$, and declines slowly at higher eccentricities. A steep decline in density is also observed toward the fovea, culminating in a rod-free zone representing approximately $1°$ of visual field. Cone density is highest in the central fovea, where it may vary by an order of magnitude across individuals to a maximum of approximately $300,000 / mm^2$, and declines rapidly at higher eccentricities. The vast majority of cones are either L- or M-cones (approximately 90% of all cones), with a 1:1 ratio across the retina. S-cones are practically absent in the foveola in accordance with psychophysical data indicating that the center of the fovea is insensitive to blue light. The two classes of photoreceptors differ in the pattern of their synaptic connections with other retinal cells. Rod terminals, or spherules, contain synapses formed by the dendritic tips of up to three rod bipolar cells and the axonal tips of H1 horizontal cells (see below). These synapses are located within a single invagination of the rod spherule. Cone terminals, or pedicles, are much larger in order to accommodate dozens of synapses with bipolar and horizontal cells. Cone-bipolar synapses are either invaginating or flat. The former are located in small invaginations of the pedicle membrane and are arranged in triads, each consisting of the tips of two horizontal cell dendrites that flank the tip of a single bipolar cell dendrite. In the primate retina, each cone pedicle contains as many as 25 invaginations allowing for multiple synaptic contacts with the same bipolar cell. In addition, each cone makes contact with a second bipolar cell at the base of its pedicle (flat contact). Cones are electrically coupled, through gap junctions, with other cones and with rods. Photoreceptors respond to light with a graded hyperpolarization: neither rods nor cones generate action potentials.

There are three morphologically distinct classes of *bipolar* cells in the primate retina: midget and diffuse bipolar cells receive input exclusively from cones, whereas rod bipolar cells receive exclusive input from rod photoreceptors. In the central retina, midget bipolar cells receive input from a single cone with either L- or M- wavelength sensitivity. There are two subtypes of midget bipolar cells: flat midget bipolar cells, that contact cones at the base of their pedicle, and invaginating bipolar cells. Diffuse bipolar cells receive input from many cones, regardless of wavelength sensitivity. Diffuse bipolar type

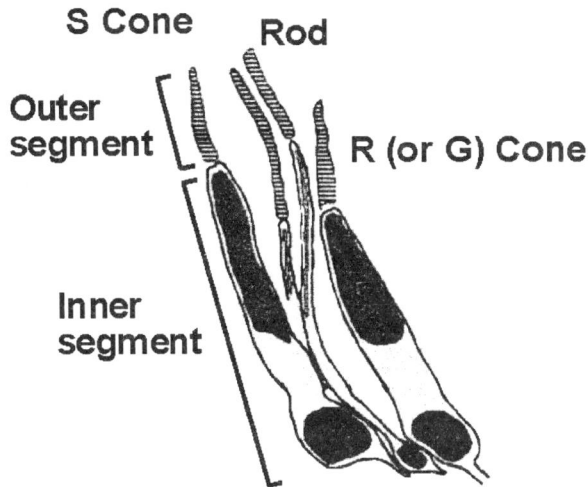

FIGURE 24.
Schematic representation of the three types of human photoreceptors: two rods, an R cone and an S-cone are shown. (From Ahnet, 1998, *Eye*, 12. Copyright 1998 by the Royal College of Ophthalmologists).

cells 1, 2, and 3 contact cones on the pedicle base (flat synapse), whereas cell types 4, 5, and 6 form invaginating synapses with cones.

Yet another type of bipolar cell receives exclusive input from, possibly, several cones with maximal sensitivity to short wavelengths (S-cone bipolar cell). Finally, rod bipolar cells can be distinguished from cone bipolar cells on the basis of the morphology of their dendritic tree and their synapses with rods. It is estimated that in the central retina rod bipolar cells represent approximately 20% of the population of bipolar cells, each receiving input from several rods.

On the basis of their response to light, bipolar cells are also classified as ON-center or OFF-center (this distinction will be discussed in more detail in the next section). Flat midget cone bipolar cells respond to a spot of light, directed to their receptive field center, with a graded hyperpolarizing response (i.e., they are turned off by light and are called OFF-center cells), whereas invaginating cells depolarize (i.e., they are turned on) in response to light at the center of their receptive field (ON-center cells). Diffuse bipolar cells type 1, 2, and 3 are OFF-center cells, whereas diffuse cells type 4, 5, and 6 produce ON, or depolarizing responses to a small spot of light. As explained in more detail below, bipolar cells provide excitatory input to retinal ganglion cells probably by using glutamate as a neurotransmitter. As shown in the simplified wiring diagram in Figure 25, ON-bipolar cells synapse with ganglion cells that have dendrites in the inner half of the inner plexiform layer, whereas the contacts formed by OFF-bipolar cells on ganglion cells are located in the outer portion of the inner plexiform layer. The synaptic pattern of rod bi-

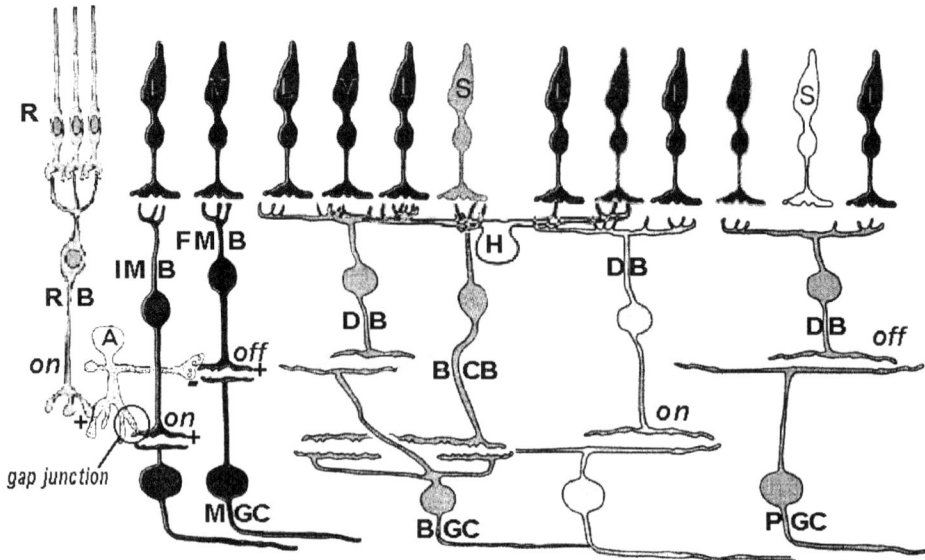

FIGURE 25.
Schematic drawing of the main cell types and their connections in the macaque fovea. The photoreceptor layer contains rods (R), long-, medium, and short-wavelength cones (L - red, M - green, or
S - blue, respectively). Invaginating (IMB) and flat (FMB) midget bipolar cells receive input from
either a red or a green cone but each cone normally provides input to both types of bipolar cells
(connections not shown here for simplicity). IMB cells are depolarized by light (ON response) and
synapse with midget ganglion cells (MGC) in the inner part of the inner plexiform layer, whereas
FMB cells display hyperpolarizing (OFF) responses to light and synapse with MGCs in the outer
portion of the layer. Diffuse bipolar cells (DB), like horizontal cells (H), receive input from all
three types of cones. However, the H1 cell displayed here receives very little input from S-cones.
One class of OFF-center DB cells makes excitatory synapses with the small bistratified ganglion
cells (BGC). This pathway accounts for the BGC's hyperpolarizing (OFF) response to mixed green
and red light. The same BGC receives excitatory input from S-cones via a blue cone bipolar cell
(BCB). This pathway accounts for the BGC's ON response to blue light. Parasol ganglion cells
(PGC) receive input predominantly from DB cells, which are classified as ON- or OFF-center depending upon the portion of the inner plexiform lamina where they synapse with PGCs. Only one
of the many functions of amacrine cells (A) is shown: it conveys signals from rod bipolar cells
(RB) via electrical synapses to MCG cells (sign-conserving synapse indicated by the "+" sign) and
transfers the same rod signal to FMB and MCG cells inverted (indicated by the "-" sign). Thus,
light-induced depolarization of RB cells depolarizes ON MCG cells and hyperpolarizes OFF MCG
cells. (From Martin, 1998, *Journal of Physiology,* 513. Copyright 1998 by The Physiological Society).

polar cells is fundamentally different: it appears that they provide input to retinal ganglion cells only indirectly through amacrine cells. Rod bipolar cells respond to light with
graded depolarization and can, therefore, be classified as ON-center cells.

The ON-center signal is transmitted to retinal ganglion cells via an electrical synapse
with an ON-center cone bipolar cell. In this pathway, light hyperpolarizes rod photoreceptors which in turn depolarize rod bipolar cells. The latter provide depolarizing input to
amacrine cells that is passed on to ON-center cone bipolar cells. The excitatory synapse
between the ON-center bipolar and an ON-center ganglion cell ensures that the latter
depolarize in response to light captured by rod photoreceptors. Stimulation of rods activates OFF-center retinal ganglion cells through the following mechanism: depolarization
of rod bipolar cells is conveyed to amacrine cells via an excitatory synapse. Amacrine
cells provide, in turn, inhibitory (i.e., hyperpolarizing) synaptic input to OFF-center bipolar and ganglion cells. In this way, light captured by rods depolarizes ON-center ganglion cells and hyperpolarizes OFF-center ganglion cells (Wässle et al., 1991).

Two types of *horizontal* cells have been identified in the outer plexiform layer of the
primate retina, H1 and H2 cells. While H1 cells form contacts with both rods and cones,
H2 cells synapse only with cones There is some evidence that the axon and the dendritic
tree of H1 cells form two distinct modules, the former making exclusive connections
with rods, the latter synapsing with cones. H1 cells, which correspond to B cells in other
mammals, are very sparsely connected to S cones and receive most of their input from
both L- and M-cones. In the fovea the dendritic three of each horizontal cell extends laterally to form contacts with up to 7 cone pedicles (Boycott et al., 1987), but the number
of cones providing input to a single horizontal cell may increase to 45° in the periphery.
There is some overlap in the dendritic fields of neighboring horizontal cells so that each
cone may be contacted by the processes of up to three interneurons. H2 cells (or A cells
in other mammalian species), on the other hand, are connected to all three types of cones
(Dacey, 1996). Both types of horizontal cells display a graded hyperpolarizing response
to light. Although both types of interneurons respond indiscriminately to input from either L- or M- cones, only H2 cells hyperpolarize in response to S-cone stimulation. In
several mammalian species, the dendrites of neighboring horizontal cells of the same
type are interconnected via electrical synapses gap junctions, but it is not known if these
contacts serve a signaling (Bloomfield et al., 1995) or simply a metabolic function.

Amacrine cells are the retina's interneurons in the inner plexiform layer. They receive
photoreceptor input from bipolar cells and form a dense network of lateral contacts with
bipolar, but primarily, with retinal ganglion cells. Across species, over 40 different types
of amacrine cells have been identified, of which only a fraction have been characterized
physiologically. Certain subtypes respond to light with a graded change in membrane
potential, whereas other produce action potentials as well. AI and AII amacrine cells are
perhaps the most well described subtypes. AI possess distinctive dendritic trees and, in
addition, an axonal process. They respond to synaptic depolarization with action potentials that propagate along their axon. They produce characteristic transient responses to
light (Dacey, 1999): the same cell discharges in response to both the onset and the offset
of a light stimulus (ON-OFF responses). AII-type amacrine cells receive excitatory input
from rod bipolar cells, playing a key role in the transmission of rod signals, and also from
OFF-center midget cone bipolar cells (Grünert & Wässle, 1996). Both AI and AII cells

receive combined inputs from L- and M-cones, but lack significant input from S-cones (Dacey, 1999). Amacrine cells employ several neurotransmitters, including GABA, glycine, and acetylocholine -- often in the same cell. Amacrine cells also make synapses with every type of retinal neuron (bipolar, ganglion cells, and other amacrine cells) and, by virtue of a long axonal process that is found in certain subtypes, have the capacity to regulate synaptic flow across retinal layers and over long distances within the inner plexiform layer. These features suggest a role for amacrine cells as the primary modulators of neuronal signaling in the retina.

Finally, primate *ganglion* cells can be classified morphologically into three types. Midget (midget or B) cells have small dendritic fields, receive input from L- and M-midget bipolar cells, and project exclusively to the parvocellular laminae of the dorsal lateral geniculate nucleus (see Figure 26). ON-bipolar cells provide input to ON-ganglion cells, whereas OFF-retinal ganglion cells are connected to OFF-bipolar cells. Midget ganglion cells comprise approximately 75-80% of all retinal ganglion cells in the central 1° of the fovea. In this region each midget cell receives input from a single ON- or OFF-bipolar cell. Another class of retinal ganglion cells are parasol cells (parasol or A cells). They are characterized by large somata and dendritic fields, receive input primarily from diffuse bipolar cells (Jacoby et al., 2000), and project exclusively to the magnocellular laminae of the dLGN. A third class of morphologically distinct retinal ganglion cells are the small-field bistratified cells, named to reflect the fact that their dendrites receive synapses from bipolar cells in both the inner and the outer half of the inner plexiform layer.

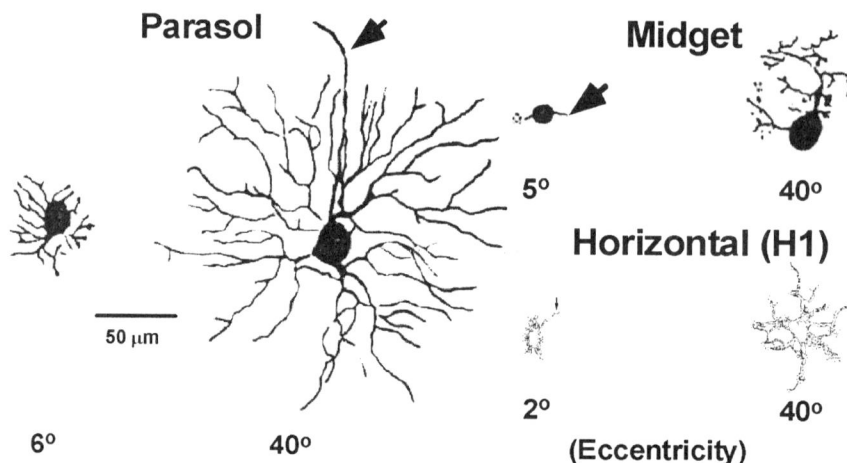

FIGURE 26.
Drawings of retinal neurons at different eccentricities. The arrows indicate axons. (Reprinted from Trends in Neurosciences, Shapley & Perry, 1986, Cat and monkey retinal ganglion cells and their visual functional roles, pp. 229-235, Copyright 1986, with permission from Elsevier Science; and from Wässle, Boycott, & Peichl, *European Journal of Neuroscience*, 1, pp. 421-435, Copyright 1991, with permission from Blackwell Science Ltd).

These cells receive excitatory input from S-cone bipolar cells that accounts for their depolarizing response to stimulation by blue light. They also receive excitatory input from OFF-center diffuse bipolar cells which, as mentioned previously, receive combined input from L-and M-cones. This unique synaptic pattern seems to account for the bistratified cell's OFF-responses to yellow light.

Bistratified cells project to two of the newly identified koniocellular layers, a population of cells located between the main geniculate layers (White et al., 1998; for a review, see Hendry & Reid, 2000[3]).

Ganglion cells are the only cell types in the retina, along with certain amacrine cells, that respond to light with action potentials that propagate along their long axons (optic nerve fibers). Typically, a ganglion cell's response to a visual stimulus is measured by its rate of firing, e.g., the number of action potentials recorded after the onset of the stimulus, in the unit of time. Until recently, it was assumed that describing the visual responses of individual retinal ganglion cells was sufficient to provide an accurate picture of the representation of any visual stimulus created inside the retina.

However, reports that the firing patterns of neighboring ganglion cells are intercorrelated, raised the possibility that visual information is transmitted to the brain, at least in part, encoded in the relative timing of spike trains from a number of optic nerve fibers (Meister, 1996).

Questions have also been raised regarding the validity of the average firing rate as the primary retinal code. This is the only type of information from the retina that is used by the brain to accomplish every visually guided behavior. In search of alternatives, researchers are now examining more closely the morphology and time evolution of the train of spikes recorded from each cell after the presentation of a visual stimulus. According to some reports, the temporal layout of individual components of this response, in the form of distinct firing event, may convey information about the properties of the stimulus (Berry et al., 1997).

In summary, light hyperpolarizes photoreceptors reducing the rate of glutamate release. This in turn depolarizes ON-bipolar cells but hyperpolarizes OFF-bipolar cells. In the former, this leads to an increase in the release of an excitatory neurotransmitter, probably glutamate, on ON-retinal ganglion cells. Conversely, OFF-bipolar cells are depolarized when the light stimulus is switched off, releasing an excitatory neurotransmitter, which again, depolarizes OFF-ganglion cells. Within each retinal layer local, circuits are formed that allow for lateral interactions between adjacent retinal neurons, mediated by horizontal and amacrine cells. These circuits enable the responses of bipolar or ganglion cells to be influenced by the responses of adjacent cells that sample neighboring portions of the visual array.

The retina: Contributions to vision

Already at the level of the photoreceptor-bipolar synapse, the neuronal signals elicited by the simplest light-dark transition have been "reshaped" as a result of the response characteristics of the photoreceptors, bipolar, and horizontal cells. These signals are affected by the activity history of each photoreceptor cell, overall levels of illumination, and the distribution of light levels across specific parts of the retina. Complex interactions that occur at every level in the retina, mainly among sensory neurons and interneurons, determine the final form of the neuronal signal that is sent to the brain. In this sense, retinal processes play a key role in several important sensory phenomena like visual adaptation, spatial and temporal resolution, and wavelength discrimination.

Visual adaptation

Adaptation is largely responsible for the ability of the visual system to retain adequate sensitivity over a wide range of light intensities. Indeed, detection of light intensity increments and decrements is preserved despite variations in the level of background luminance in the order of 10^8. Adaptation is accomplished by peripheral, receptoral, and neural mechanisms. In an attempt to describe the role of each mechanism, scientists typically compare the adaptation properties of individual components of the visual system (for instance, retinal ganglion cells) with measures of how the visual system as a whole performs under similar conditions. Adaptation experiments with humans or behaving animals typically involve determination of the visual threshold (i.e., the smallest intensity at which a light stimulus can be detected) under different levels of background illumination. "Light adaptation" protocols examine the effects of background lights on visual sensitivity. A consistent finding in these experiments is that the visual threshold increases as the intensity of the background against which the stimulus is presented is elevated. Desensitization of the visual system, which occurs very rapidly (within 1-2 seconds), is counteracted by adaptation mechanisms as discussed below.

Peripheral mechanisms involve changes in the aperture of the eye, through the constriction and dilation of the pupil, to adjust the amount of light that reaches the retina. Given that changes in pupil diameter can account for only a small portion of the dynamic range of the human visual system, processes that take place at the receptor level and beyond must be primarily responsible for adaptation phenomena. Photoreceptors play a key role in sensory adaptation through a variety of mechanisms. First, the mere presence of two types of photoreceptors is sufficient to extend the dynamic range of the visual system by several orders of magnitude. Rods are highly sensitive, capable under certain circumstances to produce an electrical response upon absorbing a single photon, and operate primarily under low levels of ambient illumination (scotopic levels). At higher levels of illumination (photopic range), cones dominate the transmission of retinal images with a high degree of fidelity. Second, individual photoreceptors can adapt to background levels of illumination, maintaining their ability to respond to small changes in the intensity of a visual stimulus presented against a wide range of background intensities. Biochemical control of the phototransduction process is primarily responsible for counteracting the effects of (background) light and is mediated by at least three calcium-dependent processes (for reviews, see Perlman & Normann, 1998; Koutalos & Yau, 1996). Specifically, inactivation of cGMP associated with light stimulation is counteracted by increased activity of the enzyme that synthesizes cGMP. This enzyme becomes reactivated when intracellular Ca^{2+} levels drop. The intracellular levels of cGMP are regulated by a second process as well: light-induced reduction in Ca^{2+} concentration suppresses the activity of the enzyme that inactivates cGMP. Again this process rapidly restores cGMP levels, at least partially, and has the tendency to reverse light-induced membrane hyperpolarization.

BOX 1. ASSESSMENT OF RETINAL FUNCTION:
THE ELECTRORETINOGRAM

The collective response of large numbers of retinal cells in response to visual stimulation can be detected noninvasively by recording the electroretinogram (ERG). The recording is performed with an electrode embedded into a corneal contact lens referenced to a skin lead (usually on the forehead or orbital rim) or to a second electrode that makes contact with the conjuctiva. Standard clinical applications of the ERG involve stimulation with brief light flashes. While certain protocols evaluate the combined response of both types of photoreceptors, the electrical responses of cones or rods can be recorded in isolation (Marmor et al., 1991).

Stimulation with a light flash after several minutes of staying in the dark (dark adaptation) produces an ERG response consisting of three major deflections (known as the a-, b-, and c- waves). The early waves peak within 31-33 ms after stimulus onset, whereas the c-wave is a slow potential that lasts for several seconds.

Most of the voltage fluctuations reflected in the ERG appear to be the result of light-induced changes in non-neural retinal cells, predominantly Müller and pigment epithelial cells. Specifically, it is believed that the b-wave is generated as large numbers of Müller cells depolarize in response to the influx of K+ from the extracellular space, the concentration of which fluctuates as a result of light-evoked changes in the activity of neural retinal elements. The c-wave is generated predominantly in the pigment epithelium as a result of photoreceptor activity. The a-wave which precedes the b-wave reflects primarily light-induced electrical changes in the photoreceptors. High frequency oscillatory responses can also be detected during the course of the b-wave, possibly reflecting the depolarizing responses of amacrine cells. ERG measurements (often in combination with other tests of visual function, such as Visual Evoked Potentials) can greatly aid the diagnosis of retinal disease such as congenital night blindness, and cone degeneration and help determine the cause of certain pathological conditions such as optic atrophy.

Another consequence of reduced intracellular Ca^{2+} concentration during phototransduction is an increase in the affinity of cation channels for cGMP. Thus, although cGMP levels fall rapidly in response to light stimulation, the remaining cGMP is sufficient to reopen some of the ion channels and partially repolarize the cell.

Contrast enhancement and visual resolution
Bipolar cells respond to photoreceptor hyperpolarization in one of two ways: one class of cells (flat midget cells) hyperpolarize, while a second class of cells (invaginating midget

cells) display depolarizing responses. Given that each photoreceptor is connected to both types of bipolar cells, the same stimulus generates two opposing responses in distinct neuronal populations in the retina. Whereas ON-cells depolarize when a light spot is turned on, OFF-cells depolarize when the light stimulus is turned off. Another characteristic of bipolar cells is that their response to a small light stimulus varies with its precise location and spatial configuration. Thus, bipolar cells depolarize when a small spot of light is flashed in one location and hyperpolarize when the same stimulus is flashed on an adjacent spot. In other words, bipolar cells show spatially distinct receptive fields. The receptive field[4] of most bipolar cells (including midget, diffuse, and S-cone bipolars) is concentrically organized with the center and the annular periphery (surround) displaying antagonistic sensitivity to light (see Figure 27). One class of cells shows an increase in their firing rate when a spot of light is flashed at the center of their receptive field, and a reduction in activity when light falls in the surround, hence the name ON-center/OFF-surround (or simply ON) cells. ON-cells, then, respond optimally to a small spot that is brighter than the background (or, alternatively, when the background illumination is turned off). Other cells increase their firing rate when stimulated by light in the surround region or, alternatively, when the receptive field center is darkened.

These OFF-center/ON-surround (or simply OFF-) cells are sensitive to spots which are darker than their background and also respond when a central light spot is turned off.

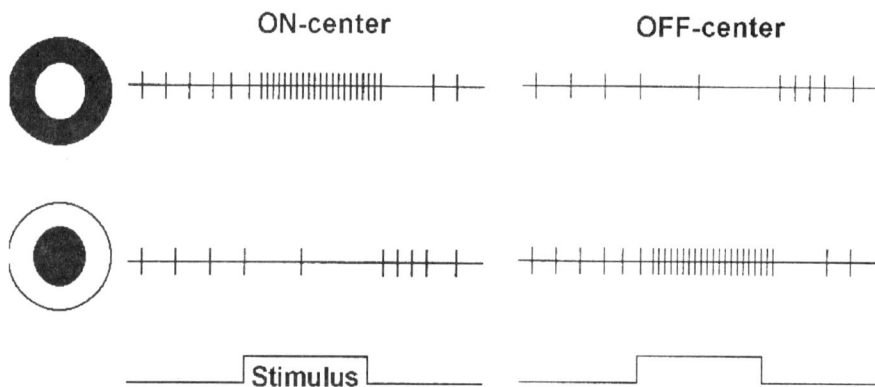

FIGURE 27.

(Left) Response profile of a prototypical ON-center ganglion cell. An increase in the illumination of the center of the receptive field is associated with an increase in the cell's firing rate above background (i.e., prestimulus) levels. In this case, the response is tonic -- that is, the train of action potentials persists throughout the duration of the stimulus. If the illumination of the annular surround area is instead increased, the cell's firing rate is reduced below background levels. A brisk response is observed when the surround stimulation is turned off. Finally, simultaneous stimulation of the center and the surround regions results in a very weak response, reflecting the antagonistic integration of inputs from the two regions. (Right) The response profile of a prototypical OFF-center cell is a mirror image of the profile of the ON-center cell.

The response of bipolar cells to light stimulation of their receptive field center (and the ON-/OFF-dichotomy) is determined directly at the photoreceptor-bipolar cell synapse. In both types of bipolar cells, signals transmitted by photoreceptor cells are conveyed by glutamate, which binds on two different types of receptors (Euler et al., 1996). ON bipolar cells contain glutamate receptors that are sensitive to the synthetic glutamate agonist amino-4-phosphonobutyric acid (AP-4). Activation of these receptors by glutamate, in darkness, closes non-specific cation channels hyperpolarizing the cell membrane. A reduction in glutamate release triggered by light opens the cation channels and depolarizes the cell. Conversely, OFF bipolar cells bind glutamate on receptors that are sensitive to the glutamate agonist kainate. Activation of these receptors causes opening of non-specific cation channels depolarizing the cell membrane. Such cells are depolarized in the dark in response to glutamate release from photoreceptors and hyperpolarize in response to light when glutamate release is reduced.

The response of bipolar cells to stimulation of the receptive field surround region is determined by negative feedback from horizontal cells exerted on the bipolar cell's dendrites, and on the cone pedicles that provide input to the bipolar cell receptive field center. As mentioned above, horizontal cell processes make dense contacts with cone pedicles and rod spherules located in close proximity with the photoreceptor-bipolar cell contacts. Each horizontal cell interacts with several neighboring photoreceptors. This arrangement is consistent with the proposed role of horizontal cells in regulating the cone-bipolar cell synapse through lateral inhibition. Stimulation of photoreceptors in the surround region of the bipolar cell receptive field triggers a mechanism which prevents photoreceptors that feed into the central region of the receptive field, from responding to light (Baylor et al., 1971). Horizontal cells may modulate photoreceptor membrane conductance via GABA-ergic synapses (for a review, see Kamermans & Spekreijse, 1999). Alternatively, horizontal cells may directly depolarize photoreceptor inner segments by inducing calcium-influx. The latter may be triggered directly by a neurotransmitter, such as glutamate or nitric oxide, or by local changes in the consistency of the extracellular fluid, such as changes in pH and Cl concentration. A third mechanism may involve synaptic inhibition applied directly onto bipolar cell dendrites.

The antagonistic center-surround organization[5] demonstrated by bipolar cells is essentially mirrored by retinal ganglion cells. The populations of ON-center and OFF-center cells can be thought of as forming two separate pathways, each providing the visual system with information regarding increments and reductions in brightness, respectively. Several lines of evidence suggest that the two systems remain segregated up to the striate cortex. Thus, an increase in the firing rate of an ON-center cell conveys a signal for brightness enhancement to the dLGN and, subsequently, to the visual cortex. Conversely, an increase in the activity of an OFF-center cell can be interpreted as a sign of darkness enhancement (Fiorentini et al., 1990). Each bipolar or ganglion cell with concentric antagonistic properties serves, in addition, as an elementary contrast detector. Thus, the responses of the vast majority of second- and third-order sensory neurons in the retina are optimally modulated by spatial transitions in light intensity *within* their receptive field.

ON- and OFF-center ganglion cells are evenly distributed in every part of the retina. The spacing of their receptive field centers sets the upper limit of spatial resolution for the visual system (which is known as the Nyquist limit). The Nyquist limit indicates the highest spatial frequency that can be unambiguously signaled to the central visual system. It is mathematically derived from the following formula:

$$\text{Nyquist limit} = \sqrt{3} * \alpha = 1.73 * \alpha \qquad (1)$$

At the center of the fovea, there is a one-to-one correspondence between cone receptors, midget bipolar and midget ganglion cells. Therefore, for the purpose of computing the Nyquist limit for foveal vision, the separation between neighboring retinal ganglion cells is equivalent to the intercell distance (α) between neighboring midget bipolar cells, or between cone receptors. For humans, this value is approximately 60 cycles/degree[6] for central (foveal) vision, assuming an intercell distance of 0.56 minutes of arc. The actual resolution power of the average adult human observer, measured with black and white gratings, is not significantly lower than the Nyquist limit, suggesting that at least for foveal vision, acuity is primarily constrained by the average separation of neighboring retinal ganglion cells. At the photoreceptor level, given that S-cones are virtually absent in the foveola, it is the spacing of L- and M-cones that actually determines this limit.

As previously mentioned, however, each midget cell with a concentrically organized receptive field operates, in essence, as a contrast detector. The spatial resolution of individual bipolar and retinal ganglion cells can be determined empirically by slowly sweeping black-and-white gratings with various bar widths across the neuron's receptive field while recording its firing rate. By using this procedure, researchers can determine the highest spatial frequency that is capable of eliciting a response from a given cell -- in other words, the cell's spatial resolution threshold, λ. Such measurements have revealed a linear relation between receptive field size and spatial resolution, which is illustrated in Figure 28a. This relation can be described by the following formula:

$$\lambda = 0.56 * \text{Diameter of receptive field center} \qquad (2)$$

In the fovea, the diameter of the center region of the receptive field is equal to three times the average intercell distance. Based on this estimate one can reformulate equation (2) to take into account the average distance between the sensory neurons or cones in the fovea:

$$\lambda = 0.56 * (3 \alpha) = 1.68 * \alpha \qquad (3)$$

Comparing equations (1) and (3), it is apparent that the spatial resolution limit of individual retinal ganglion cells in the fovea closely matches the Nyquist limit which, as mentioned above, is only slightly higher than the actual visual acuity of mature observers, as determined with psychophysical procedures (see Figure 28b). Given the fixed relationship between intercell distance and receptive field center size and the extensive spatial overlap of receptive fields among neighboring ganglion cells tin the fovea, it appears that receptive field size and spacing of sensory neurons in the fovea determine visual acuity to a great extent.

The relation between these factors is illustrated in Figure 29. Outside of the fovea, cone density falls steadily, and the convergence factor between cones, bipolar, and retinal ganglion cells shows a progressive increase. The density of retinal ganglion and dLGN cells also decreases as shown in Figure 30. As a result, visual acuity drops exponentially with eccentricity (see Figure 28b). It should be noted that other factors that operate beyond the retina also influence visual acuity, such as synaptic efficiency and the convergence ratio between retinal ganglion, dLGN, and cortical cells. These factors may play a greater role in determining spatial resolution during the first months of life. In addition to receptive field organization, retinal ganglion cells (and also neurons in the dLGN) can be distinguished by the temporal course of their response to visual stimuli and their sensitivity to spatial phase.

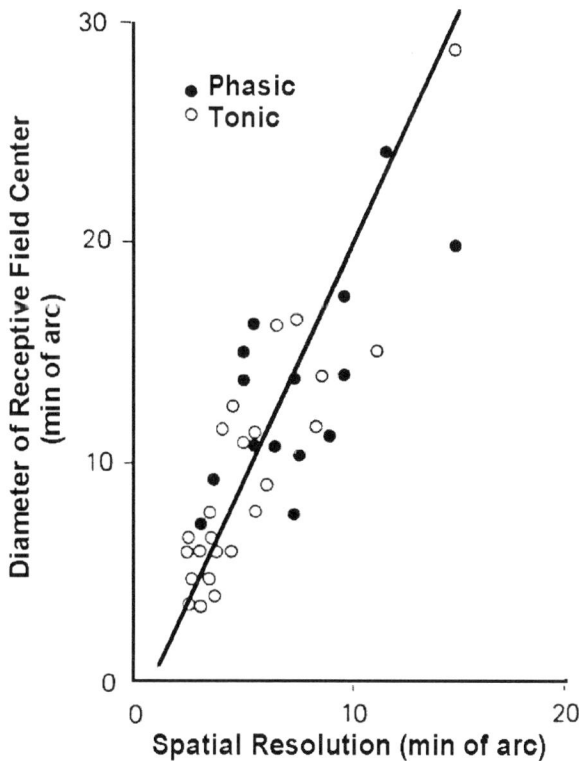

FIGURE 28a.

Estimated spatial resolution of ganglion cells in the macaque retina (deter-mined electrophysiologically as described in the text) plotted as a function of estimated diameter of the cell's receptive field center. Spatial resolution values correspond to the minimum period length of the just-resolved grating for a given cell. The slope of the regression function is 0.56, the value used in formula (2). To obtain equivalent measures in cycles/degree divide each value by 60. (From Crook et al., 1988, *Journal of Physiology*, 396. Copyright 1988 by The Physiological Society).

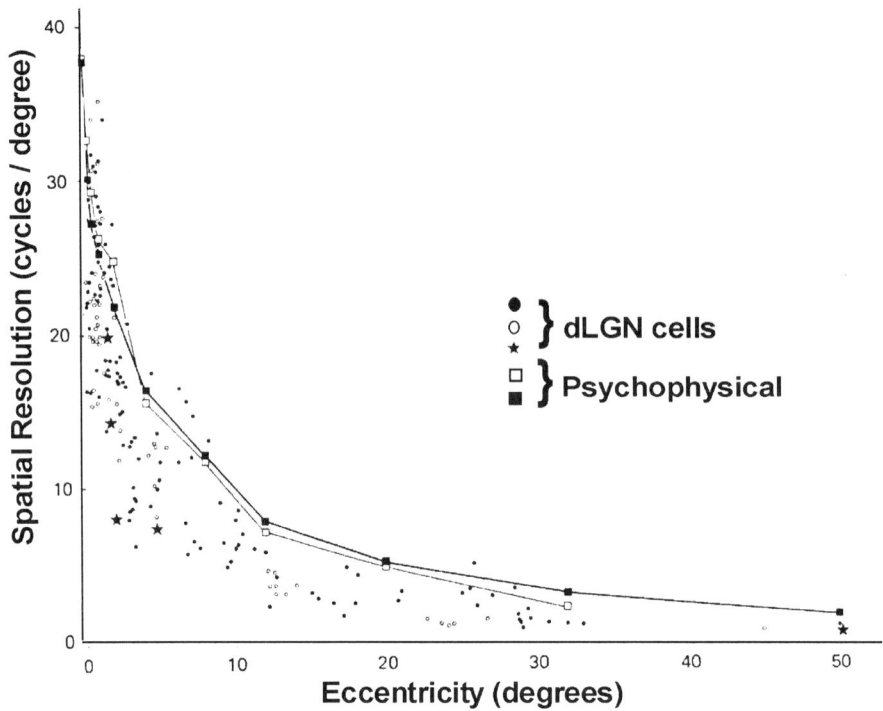

FIGURE 28b.

Plot of spatial resolution estimates obtained from individual cells in the macaque dLGN (circles
and stars) as a function of distance of their receptive field from the center of the fovea (retinal eccentricity). Visual acuity estimates obtained, at comparable contrast levels, from two human observers are also plotted for comparison. The spatial resolution of dLGN neurons is closely matched
to those of retinal ganglion cells at the same eccentricity. (From Blakemore & Vital-Durand,
1986b, *Journal of Physiology*, 380. Copyright by The Physiological Society).

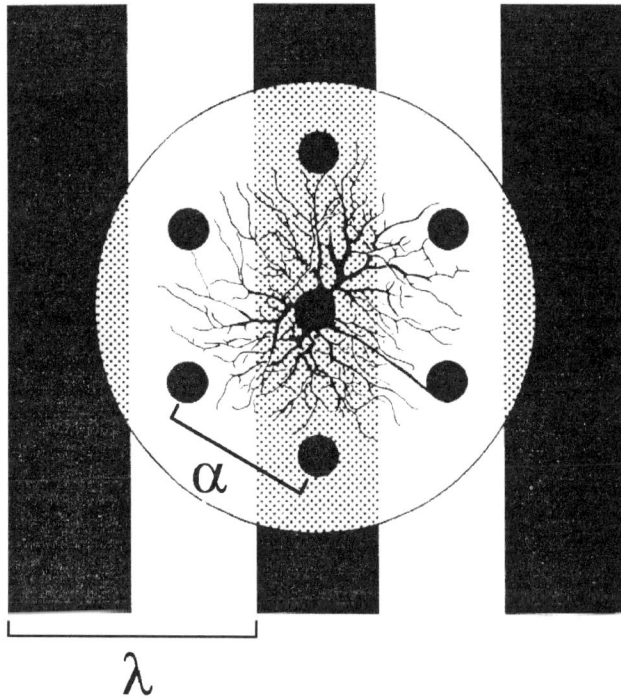

FIGURE 29.

Drawing of a cross section through the dendritic field of a midget cell superimposed on the retinal projection of a black-and-white grating. The cell's receptive field center is shown as the larger circle. The spatial frequency of the grating corresponds to the maximum spatial frequency that can be resolved by this cell. This spatial resolution limit, λ, expressed as the period length of the just-resolved grating, is larger than the diameter of the receptive field by a factor of 2. Notice also the extensive overlap between receptive fields of neighboring ganglion cells located at a distance, α, from each other. (From Wässle & Boycott, 1991, *Physiological Reviews*, 71. Copyright 1991 by the American Physiological Society).

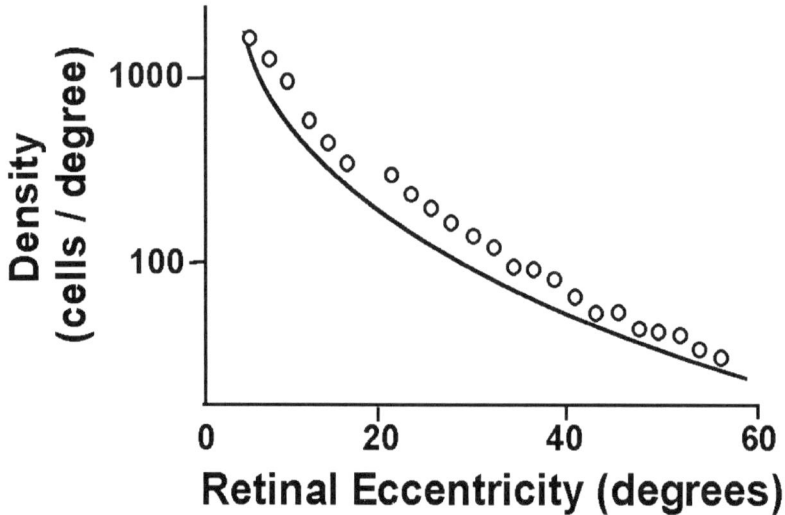

FIGURE 30.
Density of ganglion cells in the macaque retina as a function of angular distance from the center of the fovea. (From Schein & de Monasterio, 1987, *The Journal of Neuroscience*, 7. Copyright 1987 by The Society for Neuroscience).

Responses to rapidly changing stimuli

The ability to perceive rapid, local changes in brightness on the retina, produced by physical movement of elements in a visual scene or by movement of the eyes, requires retinal signals of adequate temporal resolution. One population of retinal ganglion cells is uniquely suited to transmit this kind of information to the brain with minimal distortion. Phasic cells in the primate retina, like the one displayed in Figure 31, respond briefly to the onset and offset of the stimulus regardless of its duration, do not have wavelength-antagonistic receptive fields, and probably correspond to parasol cells (Perry et al., 1984). In contrast, tonic[7] neurons produce a sustained response (i.e. the cell's firing is prolonged during the duration of the stimulus), show selectivity to stimulus wavelength, and correspond to midget cells (Perry et al., 1984). Tonic cells project to the parvocellu-lar laminae of the dLGN (i.e., laminae 3, 4, 5, and 6), whereas phasic cells project to the magnocellular laminae (i.e., laminae 1 and 2). Accordingly, tonic cells form the majority of cells in the parvocellular laminae of the dLGN in primates (P cells), whereas the ma-jority of neurons in the magnocellular laminae of the dLGN (M cells) are phasic cells. It should be noted that there is a uniform distribution of receptive field types (i.e., ON-center and OFF-center) among phasic and tonic cells.

In addition, most cells in the cat and monkey retina and dLGN can be classified as either X- or Y-like according to their spatial summation properties, which are illustrated in Figure 32 (Cleland et al., 1971; Enroth-Cugell & Robson, 1966). In the cat, tonic cells typically show linear spatial summation (see Box 2) in their receptive field (*X* cells), whereas phasic neurons display non-linear spatial summation properties (*Y* cells).

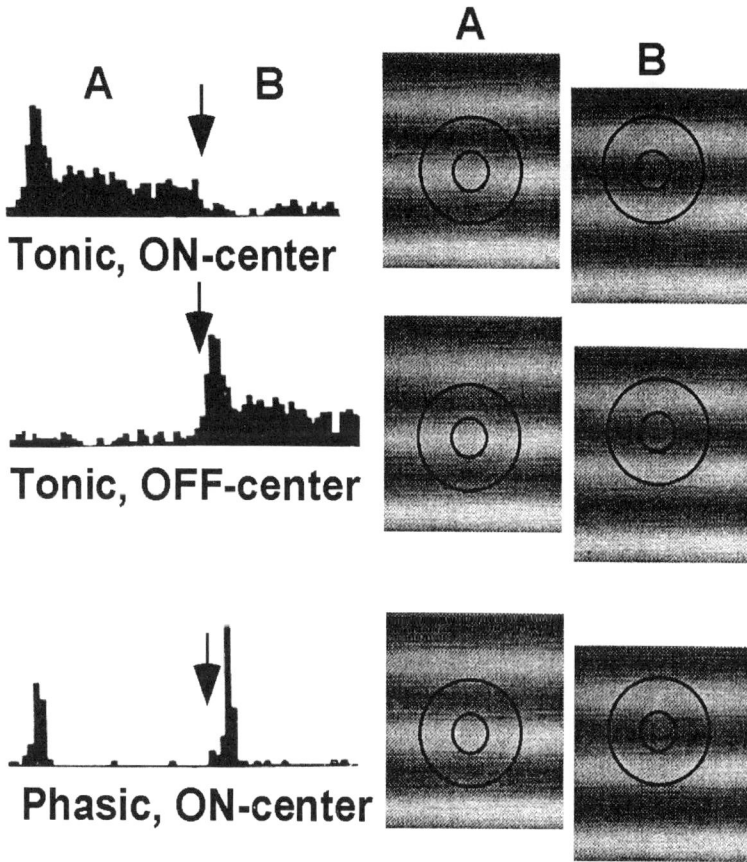

FIGURE 31.

Responses of three dLGN cells to sinusoid, black-and-white gratings shown on the right-hand portion of the figure. The cells' receptive fields are represented by two concentric circles. The gratings are phase-reversing, i.e., dark stripes are replaced by white stripes and vice versa, every 220 ms, and are positioned in such a way that the entire receptive field center (innermost circle) is either fully illuminated (A) or fully darkened (B) during the two stimulation phases. The discharge histograms displayed on the left-hand portion of the figure represent averages of the cell's response to 80 complete stimulation cycles (extracellular recordings). Contrast reversal occurred at the beginning of each epoch and at the time indicated by the arrow. The tonic, ON-center cell shows increased discharge rate throughout phase A, and returns to baseline activity during phase B (retinal ganglion and dLGN cells often show decrease in activity below baseline levels in this situation). The tonic, OFF-center cell shows the opposite response pattern: essentially no response throughout phase A, and vigorous discharge that is sustained during the entire duration of phase B. The phasic, ON-center cell responds only to the onset and offset of stimulus phase A. Macaque retinal ganglion cells show similar responses. (Based on data from Blakemore & Vital-Durand, 1986b).

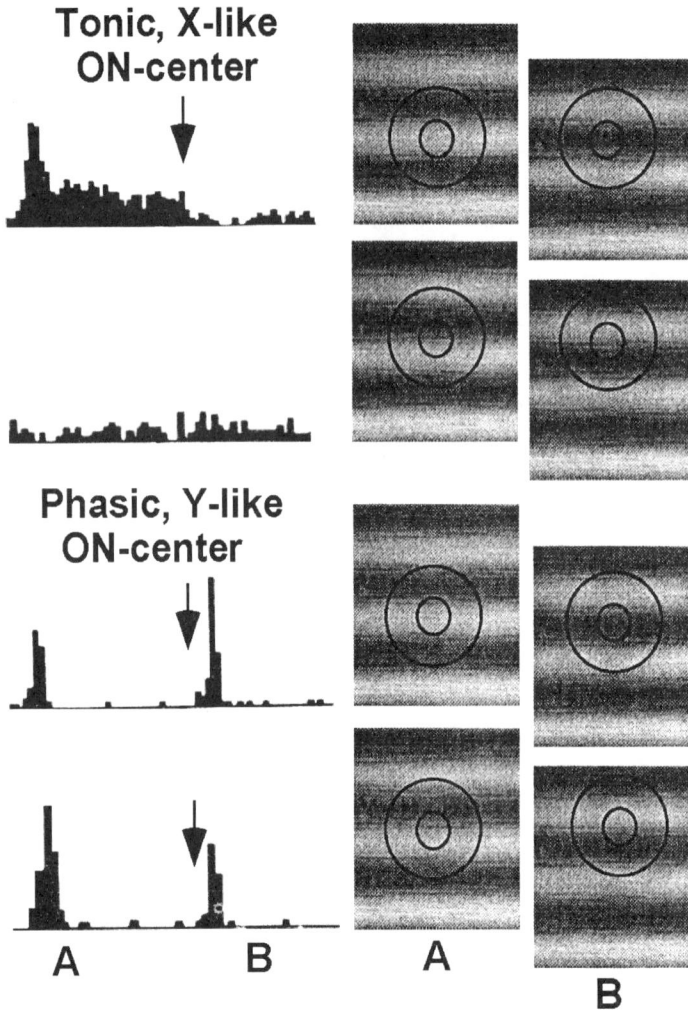

FIGURE 32.
Discharge histograms obtained in the manner described in Figure 31 from two dLGN cells, a tonic neuron with X-like spatial summation properties and a phasic neuron with Y-like summation properties. Responses were recorded during stimulus phase reversals at two grating positions. In the upper row of each set of four stimulus configurations, the center of the cell's receptive field is either fully illuminated or fully darkened as in Figure 31 (0° and 180°, respectively). The stimuli in the lower row were positioned at a phase of 90° with respect to the cell's receptive field, so that half of the receptive field was constantly illuminated. In this condition, phase-reversals were not associated with a net change in the illumination of the cells' receptive field ("null position"). The response of the tonic cell to this stimulus is not different from baseline, while the phasic cell continues to respond. (Based on data from Blakemore & Vital-Durand, 1986b).

BOX 2. MAIN RETINAL GANGLION CELL CLASSES

Midget cells	Parasol cells
Provide input to the parvocellular layers of the dLGN (P cells).	Provide input to the parvocellular layers of the dLGN (M cells).
Tonic response to step changes in illumination (X cells).	Phasic response to step changes in illumination (Y cells).
Wavelength opponent.	Do not respond to isoluminant color contrasts (Y & X cells), but may respond to color transitions independent of sign.
Linear spatial summation (X cells). (see Figure 32)	Parasol-X cells (~ 85% of total population) show linear spatial summation. Parasol-Y cells (~ 15%) show nonlinear spatial summation.
Midget-X cells: lower contrast sensitivity & shallow contrast grain function.	Parasol-X cells: higher contrast sensitivity & steep contrast grain function (X and Y cells).
Slow conduction velocity (X cells).	Fast conduction velocity (Y cells).
Small increase in dendritic field size with eccentricity (Y cells).	Steep increase of dendritic field size with eccentricity (X cells).
Slightly lower sensitivity to light than parasol cells (but also respond to scotopic light levels).	
Slightly higher spatial resolution than parasol cells (at eccentricities $> 10°$).	

In primates, Y-like cells amount for only 15% of the total number of cells with antagonistic receptive fields. In contrast with the cat, the correspondence between X and tonic (and also between Y and phasic cells) in primates is not perfect. Box 2 summarizes the most important functional properties of midget and parasol cells in the primate retinocortical pathway. When applicable, the cell class in the cat retina and dLGN that shares a particular property with one of the two primate cell classes is given in parentheses. Although the vast majority of midget cells (or P cells in the dLGN) can be classified as X-like cells on the basis of their spatial summation properties, only a small proportion of parasol cells (or M cells in the dLGN) display the non-linear summation properties of Y cells. The remaining parasol neurons show greater similarities with X cells (Blakemore & Vital-Durand, 1986b). Midget cells in primates differ from the X cells in the cat on a number of important attributes, such as contrast gain, color sensitivity, and the relation between dendritic field size and retinal eccentricity. In fact, Shapley and Perry (1986) have suggested that midget cells in primates correspond more closely to the color-selective W cells that project to layer C of the dLGN in the cat.

In conclusion, the distinction between X and Y cells in the cat retina and the dLGN does not exactly parallel the distinction between midget (tonic) and parasol (phasic) cells in primates. In the cat, transient and sustained cells differ in the size of their receptive field. Generally, transient cells have larger receptive fields (Levick, 1975). Given that a cell's ability to detect fine spatial contours, such as closely spaced black and white bars, is inversely proportional to the size of its receptive field, it was assumed that transient cells would display overall lower spatial resolution than sustained cells at the same retinal locus. This relationship has not been confirmed in primates: within the central 10° of the visual field, P and M cells in the retina (Crook et al., 1988) and the dLGN (Blakemore & Vital-Durand, 1986) have similar spatial resolution.

However, although many midget and parasol cells show similar spatial summation properties (i.e., midget-X and parasol-X cells), they do differ on a number of important properties, such as contrast sensitivity, wavelength selectivity and, by definition, the temporal evolution of their response to visual stimuli.

BOX 3. EFFECTS OF SELECTIVE LESIONS IN THE RETINOCORTICAL PATHWAYS

Parvocellular Pathway	Magnocellular Pathway
Impairment of contrast sensitivity for *high spatial* and *low temporal* frequency stimuli (Merigan & Maunsell (1993)	Impairment of contrast sensitivity for *low spatial* and *high temporal* frequency stimuli
The isolated P-pathway demonstrates poor sensitivity to low spatial and high temporal frequencies	Visibility of rapidly flickering/ moving stimuli expected to be reduced (Merigan et al., 1991)
Pronounced reduction in visual acuity	No change in visual acuity
	No change in color contrast sensitivity
Complete loss of form vision	No change in flicker resolution for high contrast stimuli

   In the dLGN, the P pathway, which is relayed via the parvocellular layers, and the M-pathway, which is relayed via the magnocellular layers, are strictly segregated anatomically. There is minimal cross-over of axonal arbors and dendritic trees between parvocellular and magnocellular layers (Michael, 1988).[8] The separation of the two pathways is maintained at least up to the afferent layer of the primary visual cortex: parvocellular afferents terminate in sublaminae IVCß and IVA, whereas magnocellular projections first synapse in lamina IVCα (Fitzpatrick et al., 1985). Further, there is evidence from the study of the effects of chemical lesions, which can be selectively targeted to either the parvocellular or the magnocellular laminae of the dLGN, that the two pathways may differ in their relative contribution to different visual functions. Some of these findings are outlined in Box 3.

Spectral opponency and color sensitivity

As mentioned earlier, in organisms that possess trichromatic vision, light that travels to the retina differentially activates three types of cones depending upon its spectral composition. Up to the level of the outer plexiform layer, signals from each cone type remain segregated as they are passed on to certain types of bipolar cells. However, cone-specific signals are inherently ambiguous: the output of each individual cone is determined by light intensity (i.e., the number of absorbed photons), as well as by the spectral composition of light (i.e., wavelength). Subsequently, cone-specific inputs are combined to generate a representation of the relative activation of different cone types by the stimulus. This code is reflected in the responses of midget and bistratified ganglion cells.

   Wavelength-specific information is conveyed primarily by midget bipolar (flat and invaginating ON-center, and flat and invaginating OFF-center cells) and midget retinal ganglion cells. In the central 50° of the retina, midget bipolar cells receive input from a single cone that shows sensitivity to either long (red) or medium (green) wavelengths. Cones that show peak sensitivity to short-wavelength, blue light, project almost exclusively to a morphologically distinct class of bipolar cells, S-cone bipolars. S-cone bipolar cells, in turn, synapse with the small-field bistratified ganglion cells. As expected from this arrangement, midget and bistratified bipolar cells show wavelength-specific responses. In contrast, diffuse bipolar and horizontal cells receive additive inputs from L- and M-cones and therefore do not contribute to the transmission of wavelength-specific information. A similar conclusion has been reached for the two most common types of amacrine cells (AI and AII).

   In the inner plexiform layer, signals from different types of cone photoreceptors and corresponding bipolar cells are combined to generate antagonistic, or opponent, receptive field properties, which are illustrated in Figure 33: each neuron becomes depolarized when the receptive field center is stimulated by light in a particular spectral band and hyperpolarized by light in a different band About half of the midget ganglion cells become depolarized by either red or green spots of light (red-ON and green-ON cells, respectively) and the remaining become hyperpolarized by the same stimuli (red-OFF and green-OFF cells, respectively). Bistratified cells become depolarized by blue light flashed at the center of their receptive field (blue-ON cells) and hyperpolarized when stimulated with yellow light. In addition, the electrical response of bipolar and retinal ganglion cells that receive cone-specific inputs is reversed when a chromatic stimulus is turned off.

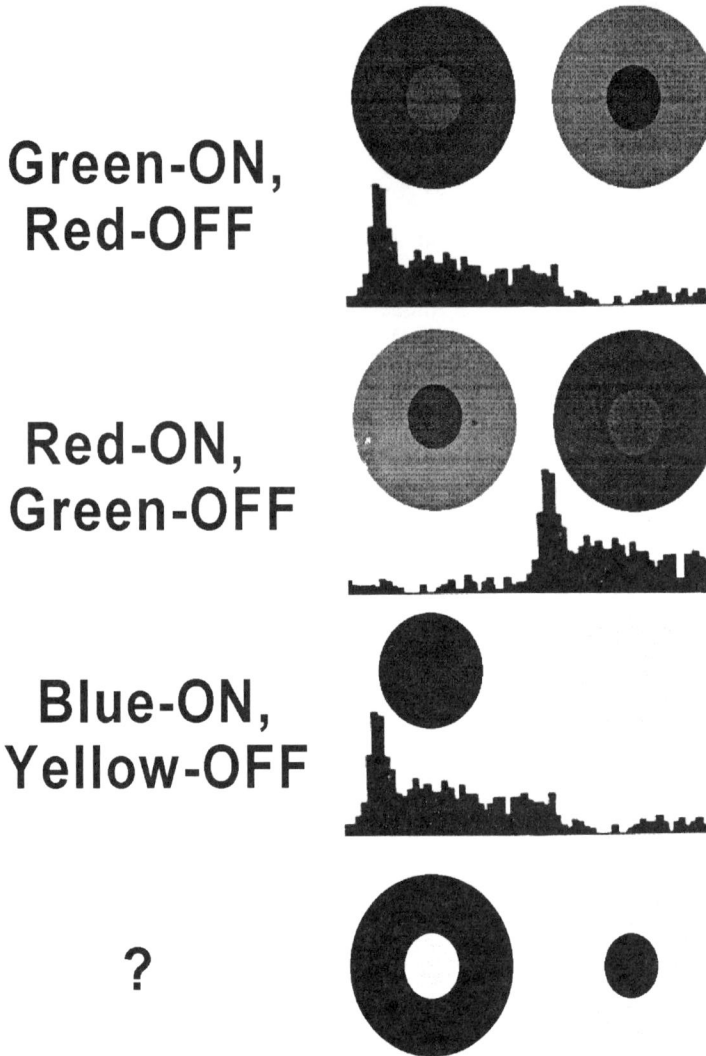

FIGURE 33.
Discharge histograms from four representative color-opponent ganglion cells in the macaque retina. The existence of receptive fields with spatially opponent concentric field organization for yellow and blue (bottom row) has not been confirmed.

For instance, a red-ON/green-OFF ganglion cell depolarizes when a green spot of
light at the center of its receptive field is turned off. Spectral opponency characterizes
cells in the central 10-degree portion of the retina. In the periphery, most midget ganglion cells receive combined input from L- and M-cones, a finding that correlates with
the documented reduction in sensitivity to hue for stimuli presented at large retinal eccentricities.

The ability to discriminate spatial variations in the spectral composition of reflected light is greatly enhanced by another functional property of retinal ganglion cells: antagonistic center-surround receptive field organization. In general terms, this implies that the spectral opponency observed for stimulation of the receptive field center is reversed when stimulation is applied in the region surrounding the center of the cell's receptive field. For instance, a red-ON/green-OFF-center ganglion cell would depolarize when a red spot of light is flashed on the receptive field center and hyperpolarize when the surround is stimulated with green light (maximal surround hyperpolarization would be obtained using an annular stimulus). As in the case of achromatic (i.e., luminance) spatial opponency, often the effect of the surround stimulation is solely to counteract the effects of center stimulation. In the previous example, this implies that green-light stimulation restricted to the surround region has no effect on the cell's membrane potential. However, when the center and the surround are stimulated simultaneously by red and green light, respectively, the cell's depolarizing response is greatly diminished or altogether abolished.

Although spectral opponency in retinal ganglion cells is known and has been studied extensively for over 30 years, the synaptic pattern that is responsible for this property is still unclear. It appears that the response profile of the receptive field center of red/green midget ganglion cells is determined primarily by the unique patterns of connections between cones, midget bipolar cells, and ganglion cells. In the central $10°$ of the retina, each midget bipolar cell receives direct input from a single L- or M-cone and each midget ganglion cell receives direct input from a single midget bipolar cell (Calkins et al., 1994).[9]

A much greater degree of uncertainty exists regarding the synaptic mechanisms responsible for the spectrally antagonistic receptive field surround in midget bipolar and ganglion cells (for reviews, see Dacey, 2000; Martin, 1998). There is consensus that, at least in the majority of midget bipolar cells, the response profile to surround stimulation is generated by inhibitory feedback from horizontal cells. However, the origin and spectral specificity of these inputs are unknown. A key issue in this proposition is that horizontal cells do not receive exclusive input from either L- or M-cones. Instead, their light responses indicate that they integrate L- and M-cone signals (Dacey et al., 1996). Accordingly, only a small minority of midget ganglion cells appear to receive exclusive input from either L- or M-cones to the receptive field surround (Smith et al., 1992). In the majority of midget cells, that receive mixed L- and M-cone input in the surround region, the relative strength of inputs from a particular type of cone to the center and the surround, may determine the cell's opponent response. This notion is based on the fact that the receptive field surround receives input by several cones, whereas the center is dominated by the input from a single cone. For instance, in the case of a red-ON/green-OFF cell, the center receives direct input from a single L-cone, whereas many L- and M-cones provide input to the surround. When red light falls exclusively on the center, the cell is strongly depolarized. However, when the surround is stimulated by green light, inputs from many M-cones activate horizontal cells which in turn hyperpolarize their bipolar cell target.

The picture is slightly clearer with respect to the processes involved in shaping the responses of the blue-ON/yellow-OFF small bistratified ganglion cell. These cells receive direct excitatory input from S-cone bipolar cells, which already display blue/yellow, center/surround, antagonistic receptive fields. In these cells, depolarizing input from S-cones is responsible for the blue-ON receptive field center. Whereas, combined hyperpolarizing input from L- and M-cones is probably responsible for the antagonistic surround. In addition, bistratified cells receive hyperpolarizing input from diffuse cone bipolar cells which, in turn convey combined L- and M-cone input. It is thought that the diffuse bipolar input further enhances the blue/yellow opponency (Dacey, 2000).

To summarize, light falling on every retinal location affects the activity of more than one distinct subpopulation of cells. Each class appears to be more suitable for the analysis of a different aspect of the visual array, such as contrast, color, and direction of motion. Thus, midget ganglion cells (Pß cells) are more suitable for transmitting visual details from a high-contrast visual array with a large degree of spatial resolution. Conversely, parasol ganglion cells (Pα cells) are clearly more suitable for relaying low resolution images from visual arrays that are either rapidly changing or moving at high speeds under low contrast conditions. One of the advantages of early differentiation is that a wide range of sensory cues can be extracted simultaneously from the visual input. It should be emphasized, however, that conclusions regarding the relative importance of each subcortical visual pathway in perception are difficult to substantiate given the simplicity of visual analysis that is performed at this level. In the next chapter we will give examples of how neural signals, which represent visual input, are transmitted through separate neocortical pathways or systems. According to a widely popularized notion, a more or less clear-cut correspondence exists between the subcortical afferent pathways and the extrastriate visual subsystems. At the end of this exposition, it will become evident that such simplifying notions are not adequate for describing either the complexity inherent in the organization of the visual system, or the complex nature of visual perception.

NOTES

[1] Humans and primates fixate on visual targets using the fovea, a roughly circular region with a diameter of 1-2 degrees of visual angle. In the cat, the corresponding region is known as *area centralis*.

[2] In humans, this estimate is lower, around 25%.

[3] The two koniocellular (K) layers in the dLGN that receive input from bistratified cells project to the cytochrome-oxidase rich blobs in area V1 (for more details see Chapter 5). Two other koniocellular layers contain cells with low spatial resolution that project to layer I in area V1, whereas the remaining two K layers relay visual input to the superior colliculus. These dLGN neurons are considered homologous to the W cells in the cat.

[4] Receptive field: the portion of the visual field in which a change in stimulus parameters is accompanied by a change -- either an increase or a decrease -- in the cell's electrical response.

[5] Notably, the center-surround organization of retinal ganglion cells deteriorates and, in some cases, is completely lost in dim light (Muller & Dacheaux, 1997).

[6] This value corresponds to a separation (or period) of 1 minute of arc between adjacent dark bars in a grating.

[7] Distinct transient and sustained response profiles to light stimulation have also been found in bipolar cells in non-primate species (Euler & Masland, 2000). Inhibitory feed-back from amacrine cells, exerted on bipolar cell axons, is at least partially responsible for the emergence of phasic response profiles at this level.

[8] In primates, a number of retinal M cells also project to the superior colliculus. These cells form a distinct population, which is separate from neurons that project to the dLGN (Shapley & Perry, 1986). The primate superior colliculus receives additional input from a smaller class of retinal cells, known as the "rarely encountered cells". In the cat, X and Y cells both project to layers A and A1 of the dLGN. Y cells also give out collaterals that project to the superior colliculus. W cells, on the other hand, project to layer C of the dLGN and to the superior colliculus.

[9] The response of the receptive field center of a given bipolar cell may be influenced to some extent by more than one cone (probably via electrical synapses between neighboring cones, Lee, 1996), but it is unclear how this convergence affects the spectral sensitivity of this portion of the receptive field.

Chapter 5

Cortical Visual Pathways

In addition to the existence of distinct transmission channels in the retinocortical pathway, research that started nearly three decades ago suggests that extrastriate cortical areas are specialized for analyzing different aspects of the visual input. In one of the earliest observations of anatomically dissociable functional properties in the primate visual cortex, Zeki (1971) reported that cells in the middle temporal area (MT) of the macaque monkey (located in the posterior bank of the superior temporal sulcus; see Figure 21) responded selectively to visual motion, but not to stimulus color or form. On the other hand, neurons in area V4 (located rostral to the lunate and the inferior occipital sulcus) were found to be sensitive to the color and orientation of visual stimuli, and responded indiscriminately to a large range of directions of visual motion. The idea that higher level visual processes are served by two distinct pathways was further elaborated by Ungerleider & Mishkin (1982).

At that time, it was also acknowledged that each area contained mechanisms responsible for a different level of analysis of visual input along a hierarchically organized pathway. The apparent simplicity and heuristic value of this concept triggered an impressive wave of research that resulted in the identification of several more "visual" areas. The anatomical distribution of the main extrastriate areas as they appear on the lateral surface of the primate brain can be seen in Figure 21. Each pathway originates in separate laminar subdivisions of the striate cortex (Maunsell & Newsome, 1987; Hubel & Livingstone, 1987; Livingstone & Hubel, 1987). Anatomical studies using radioactive tracers have revealed an impressive number of interconnections which, not surprisingly, run between, as well as within, the two proposed pathways (Felleman & Van Essen, 1991). Evidence for extensive convergence of inputs from different functional systems also exists. Some of the best-described connections are presented in schematic form in Figure 34. In the following sections basic information is reviewed regarding the functional and anatomical organization of these areas.

Primary visual cortex: Area V1

Intrinsic organization and circuitry

The primary visual cortex consists of six main layers that show distinct cellular composition and patterns of connectivity. Histologically, cells in the primary visual area can be classified into three broad categories that include two types of *stellate* neurons, which are multipolar cells that function as interneurons, and *pyramidal* cells. One class of stellate neurons, *spiny* cells, has visible spine-like processes on their dendrites, whereas the sec-second type, *smooth stellate* cells, lacks dendritic spines. Spiny cells, which are more numerous than smooth cells, have been labeled by various markers for the neurotransmitter glutamate and therefore, it appears that they form excitatory synapses (Freund et al., 1983). Smooth stellate cells respond positively to various markers for the inhibitory neurotransmitter GABA. Spiny stellate cells in layer IVC are the primary recipients of

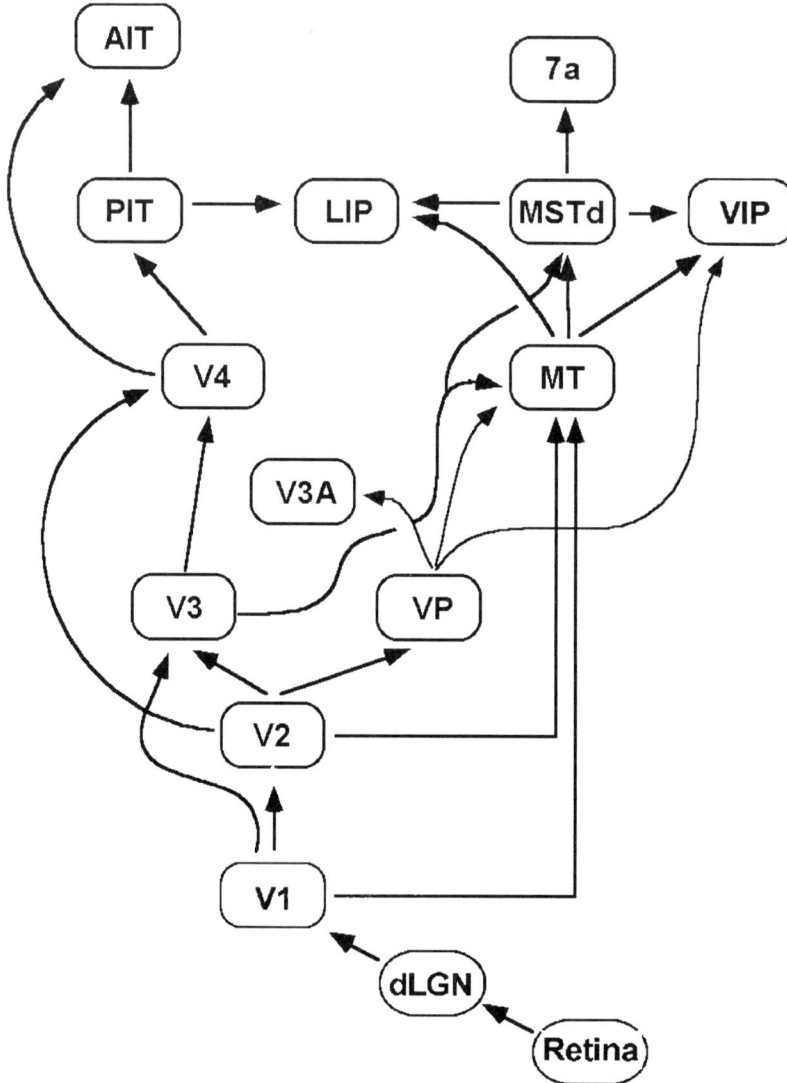

FIGURE 34.
Schematic diagram of the main connection between visual areas. Only forward connections are shown for simplicity, although the vast majority is bi-directional.

visual thalamic input (McGuire et al., 1984).[1] Neurons in this layer have concentric, antagonistic receptive fields very similar to those found in retinal ganglion cells and in the dLGN. It has been estimated that between 400 and 800 geniculocortical fibers cross any given point in layer IV (Freund et al., 1985), suggesting that there is an extensive overlap among the terminal fields of single afferents.

In their attempt to understand the contribution of a given cortical area to vision, physiologists utilize three kinds of information: cortical circuitry, regional functional differentiation, and the responses of individual neurons to visual stimulation. This section focuses on cortical circuitry in the primary visual area. The other two aspects of neuronal organization will be examined in subsequent sections. *Cortical circuitry* refers to the pattern of intrinsic connections that exist between cells in one area, and to the organization of connections formed with other brain areas. The primary visual cortex contains an abundance of horizontal, intralaminar connections, and of vertical, interlaminar connections, as illustrated in Figure 35. These connections account for a significant proportion of the excitatory input to V1 neurons, which ranges between 50 and 70% (Chung & Ferster, 1998). The remaining excitatory drive is provided by direct thalamic input. Layers IVCß and IVA receive the majority of afferents from the parvocellular layers of dLGN.[2] Afferent signals, relayed via the spiny-dendrite cells in IVCß are directed predominantly to laminae IIIB and IVA (Fitzpatrick et al., 1985). Smooth stellate cells in IVCß project to laminae IVA, VI, IIIB, and also to lamina IVCa -- the primary target of magnocellular Input from the dLGN.

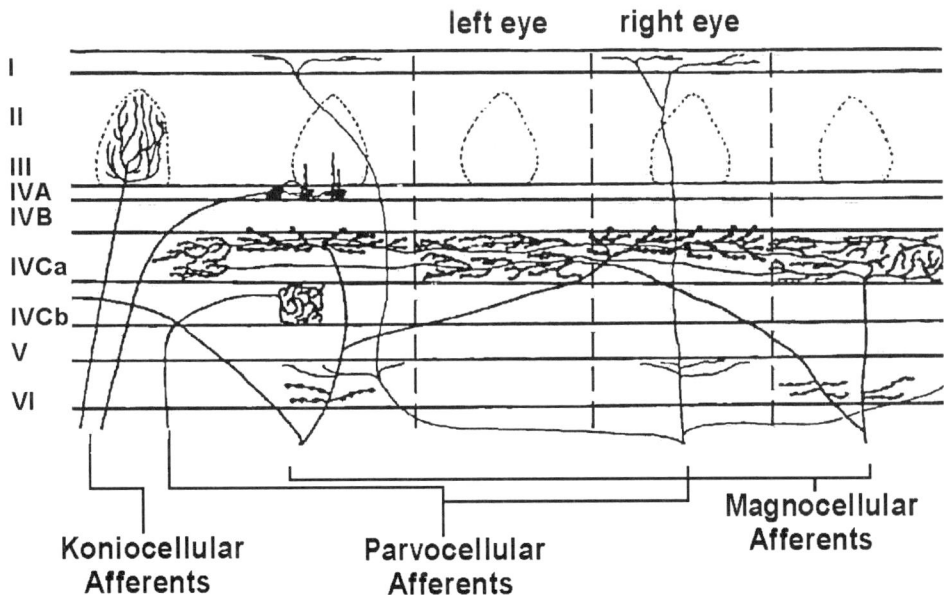

FIGURE 35.
Schematic rendering of the main branching patterns formed by geniculocortical afferents within different laminae in area V1 in the macaque. (From Fitzpatrick et al., 1985, *The Journal of Neuroscience, 5*(12). Copyright 1985 by The Society for Neuroscience).

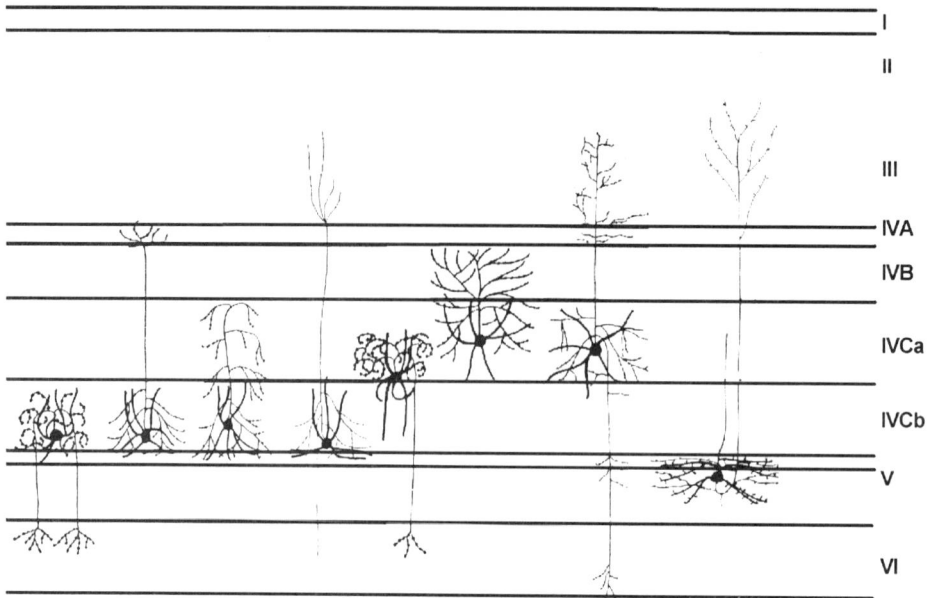

FIGURE 36.

Schematic rendering of the interlaminar branching patterns of local circuit neurons in the macaque area V1. (Based on data from Lund, 1987).

As shown in Figure 36, the bulk of the vertical projections from lamina IVCa are directed to lamina IVB (Fitzpatrick et al., 1985). The latter mainly forms connections with extrastriate areas including areas V2, V3, and MT. These projections are made by pyramidal neurons that have long dendritic and axonal processes and form excitatory synapses. Finally, the supragranular layer IIIB receives its primary input from layers IVCß and IVA, and also from the magnocellular afferent layer IVCa (Fitzpatrick et al., 1985). In general, cells in superficial layers project to other cortical areas, neurons in layer V project to the superior colliculus, whereas the feedback projections to the dLGN originate in layer VI. Given the role of the superior colliculus in the control of rapid eye movements, the cortico-tectal projection may carry signals regarding the motion of visual objects. The functional significance of the cortico-geniculate projection is not clear at present.

The afferent sublaminae differ from each other with respect to the extent of lateral connections that they contain (Fitzpatrick et al., 1985). The extent of dendritic and axonal arbors of V1 cells has been studied histologically using intracortical injections of cellular markers, such as horseradish peroxidase,[3] in order to visualize the anatomy of individual cells. The lateral spread of axons within lamina IVCa is extensive (typically in the order of 1-2 mm). The horizontal spread of axons in layer IVB is even larger (up to 4 mm). In both laminae, axons form periodic clusters of synapses with cells in the same lamina. In contrast, long-distance horizontal connections are rare in layer IVCß.

Several studies using intracellular injections of horseradish peroxidase have provided hints to the existence of complex local circuits in the striate cortex (for a review, see

Lund, 1988). These circuits appear to play an important role in modulating the input-output properties of relay cells in this area. For instance, it has been found that apical dendrites of pyramidal neurons in layer VI receive projections from cells in layer IVCa. The same cells in layer VI send axon collaterals to layer IVCa cells (which terminate primarily on neurons with smooth dendrites). Thus, a given pyramidal cell in layer VI may receive thalamic input both directly (via collateral branches of magnocellular or parvocellular thalamocortical afferents) and indirectly (through lamina IVCa). Layer VI cells may modulate the activity of neurons in layer IV via axo-axonic synapses. Smooth stellate neurons in layers VA and IVA also form local circuits. Cells in these layers form connections with neurons in afferent layers (IVCa and IVCß), which are presumably inhibitory. Layers VA and IVA are also reciprocally interconnected. It is hypothesized that the role of efferents from layer VA is to gate the transmission of thalamocortical input by neurons in the afferent layers.

In addition to its organization into horizontal layers, the primary visual cortex in primates is organized into vertical patches or columns on the basis of both functional (i.e., stimulus preferences) and anatomical characteristics (i.e., pattern of connections). The most well-established vertical differentiation involves what is known as ocular dominance columns. Cells in layer IVC receive inputs from a single eye. This layer is segregated into alternating columns that contain cells which respond to stimulation of either the contralateral or the ipsilateral eye. This is established electrophysiologically by recording single unit activity along tangential electrode penetrations through layer IVC in response to monocular stimulation. The majority of cells in the other layers respond to stimulation from both eyes (binocular neurons). However, the relative contribution of each eye to the driving of individual neurons varies in an orderly fashion along the cortical surface. The distribution of ocular preferences in a population of cells can be quantified using *ocular-dominance histograms* (Wiesel & Hubel, 1963). According to this method, a neuron is classified into one of seven categories based on its relative sensitivity to stimulation of the left, the right, or both eyes simultaneously. By convention, cells in category 1 are driven exclusively by the contralateral eye, neurons in category 7 are driven exclusively by the ipsilateral eye, whereas the excitatory influence of the two eyes is balanced for cells in category 4. Cells in the intermediate classes show relative preference for one eye (see Figure 37). Neurons in regions that lie immediately above the center of an ocular dominance column in layer IVC show strong (but not exclusive) preference for one eye. Cells in the non-afferent layers that are located above the fringes of ocular dominance columns in layer IVC respond to stimulation of both eyes, but they show relative preference for either the contralateral or the ipsilateral eye. These findings suggest that projections from cells in different ocular dominance columns in layer IVC converge onto individual cells in non-afferent layers, but in most cases one set of inputs is stronger than the other. Histological cell reconstruction studies suggest that the majority of long-range intralaminar connections run between columns of similar orientation preference (Malach et al., 1993). It appears that in layer IVCß the dendritic trees of most cortical cells are restricted within same-column boundaries (Katz et al., 1989). In contrast, the terminal arbors of axons projecting from layer IVCß to layers II/III usually do not respect ocular preference column boundaries. At the macroscopic level, the existence of binocular interactions can be studied with the Visual Evoked Response method (VER). Both the initial segregation and the subsequent convergence of inputs from the two eyes -

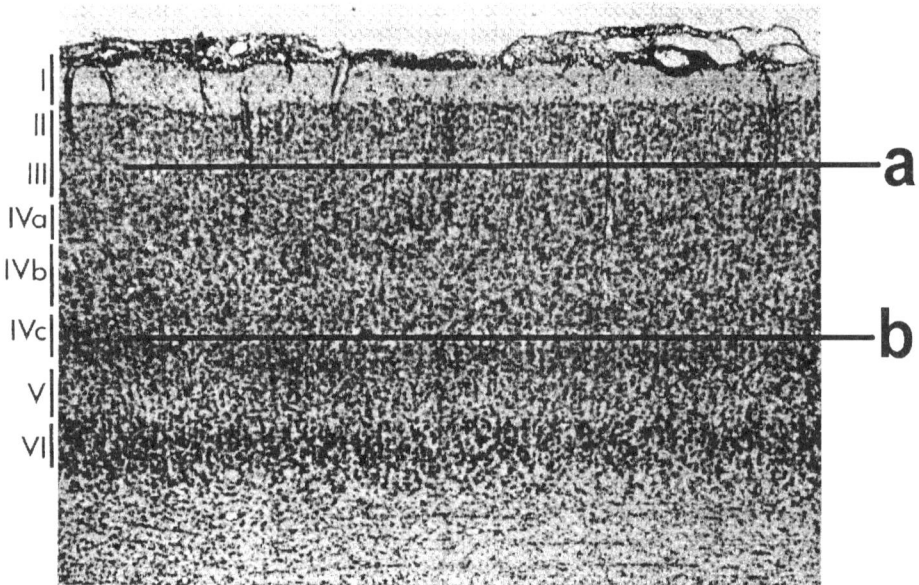

Proportion of Cells

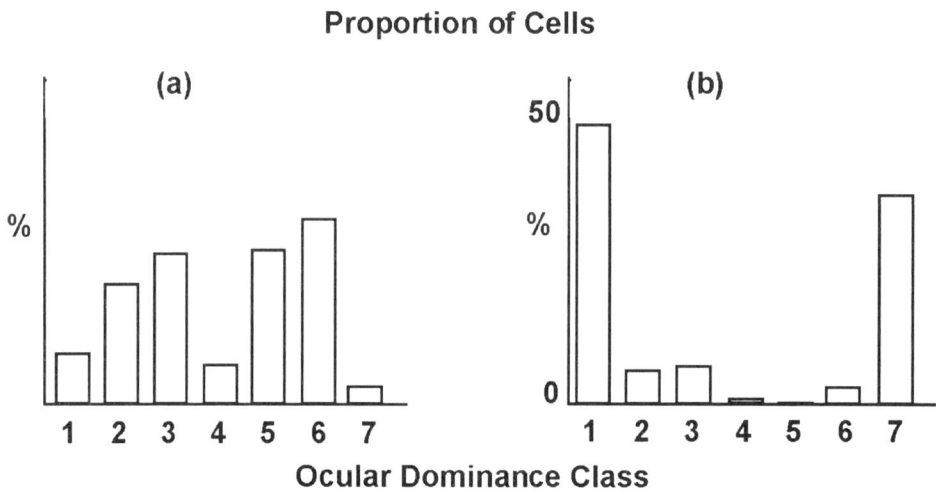

Ocular Dominance Class

FIGURE 37.
(*Top*) Radial section through area V1 showing two tangential electrode tracts. (From Laminar and columnar distribution of geniculocortical fibers in the macaque monkey. Hubel & Wiesel, 1972, *The Journal of Comparative Neurology*, 146. Copyright 1972 by John Wiley & Sons, Inc. Reprinted by permission of Wiley-Liss, Inc., a subsidiary of John Wiley & Sons, Inc.). Tract (a) traversed layer III while tract (b) was through layer IVC. (*Bottom*) Ocular preference histograms obtained from extracellular recordings along each of the two tracks. (Based on data from Hubel & Wiesel, 1968).

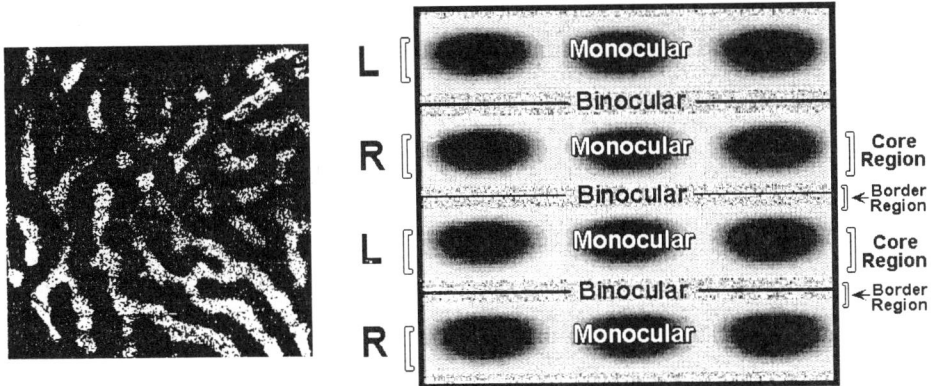

FIGURE 38.

Left: Reconstruction of tangential section through layer IV revealing cortical patches dominated by inputs from the left or the right eye using retrograde staining with [³H] Proline. Cortical patches dominated by the injected eye appear as white. (From LeVay et al., 1980). Right: Schematic drawing of the spatial relation between ocular dominance patches in layer IVC (wide horizontal strips) and blobs (dark oval shapes). Blobs are typically located within the strictly monocular core zones. Thin zones that contain neurons responsive to both eyes exist around the ocular patch boundaries. Interblob regions encompass cortex from monocular core zones as well as cortex in binocular transition zones. (From Horton & Hocking, 1998, *The Journal of Neuroscience*, 18(14). Copyright 1998 by The Society for Neuroscience).

play a crucial role in sustaining the sensitivity to binocular depth cues (i.e., stereoscopic cues).

Additional evidence for the existence of ocular dominance columns comes from histo logical studies: injection of radiolabeled amino acids (such as [³H]-Proline[4]) into one eye results in an alternating pattern of stained and unstained patches in layer IV (see Figure 38).

Cytochrome oxidase compartments

Another staining technique, using the mitochondrial enzyme *cytochrome oxidase*, reveals distinct patches of cytochrome oxidase-rich areas (*blobs*) interspersed among lightly stained areas in the non-afferent layers (known as *interblobs*; Livingstone & Hubel, 1984). It appears that blobs receive input directly from the koniocellular layers of the dLGN (Hendry & Yoshioka, 1994). Cells within the cytochrome oxidase-stained patches are more likely to show wavelength-selectivity, whereas orientation selectivity is more frequently encountered in cells in the interblob regions (Hubel & Livingstone, 1987). Neurons located within blobs typically have monocular responses. This is consistent with the observation that strands of blobs typically run down the middle of an ocular domi-nance patch as shown in Figure 38. Although cytochrome oxidase compartments may differ in the types of afferent input they receive via layer IV cells, combined cytochrome oxidase-staining and Golgi-impregnation[5] techniques revealed extensive cross-over be-tween blob and interblob regions mediated by dendritic arbors (Geyer et al., 1991, see

Figure 39). Cytochrome oxidase-staining in fixed human brain preparation results in similar tangential profiles (Burkhalter & Bernardo, 1989).

Orientation preference columns
Cells in the non-afferent layers show preference for a narrow range of orientations of line stimuli (visual edges, gratings, slits, or bars). Cells with similar orientation preferences cluster together in vertical patches of cortex known as iso-orientation columns. The existence of iso-orientation columns has been demonstrated with electrophysiological and radiographic methods. Thus, tangential or oblique electrode penetrations through area V1 reveal cells with similar orientation preferences over a distance of approximately 100 µm (see Figure 40). A shift in orientation preference in the order of 10° to 20° is noted as the electrode enters an adjacent column. A complementary method involves injecting 2-deoxyglucose, a radiolabeled form of glucose, that is the primary source of energy for nerve cells.

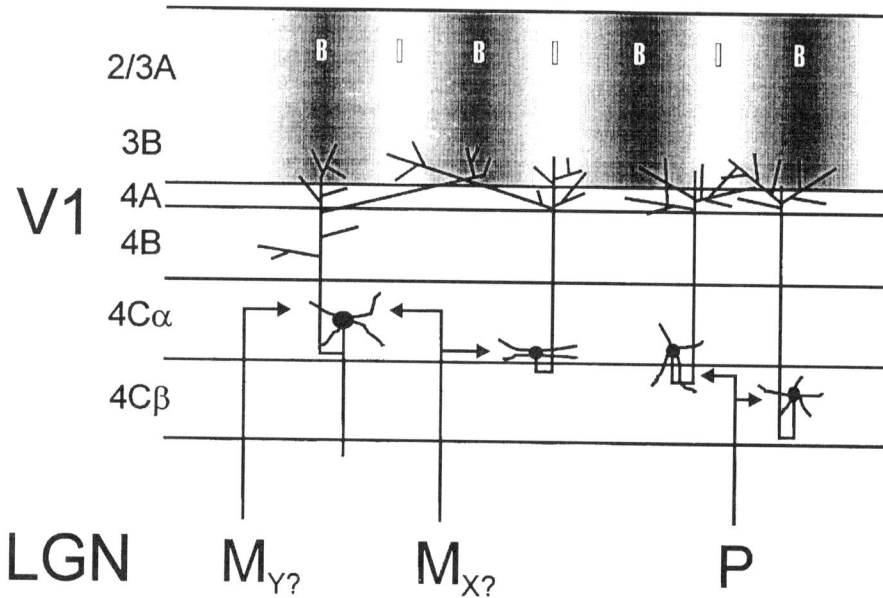

FIGURE 39.
Schematic diagram illustrating the flow of retinocortical input to blob (darker regions in layers II and III) and interblob compartments (lighter regions in layers II and III). Parvocellular afferents (P) synapse with two types of spiny stellate neurons in layer IVCβ: one type of cell (far right) has a dendritic field restricted to layer IVCβ, while the dendritic field of the second cell type often arborizes within both IVCβ and IVCα (middle right). Both types of stellate neurons project to blobs as well as interblobs. Magnocellular inputs (M) target cells with dendritic fields restricted within layer IVCα. There are indications that Y-like dLGN cells (M_Y) target spiny stellate cells which, in turn, project to blobs, whereas inputs from X-like M cells (M_X) are conveyed to both blobs and interblobs. (From Yabuta & Callaway, 1998, *The Journal of Neuroscience*, 18(22). Copyright 1998 by The Society for Neuroscience).

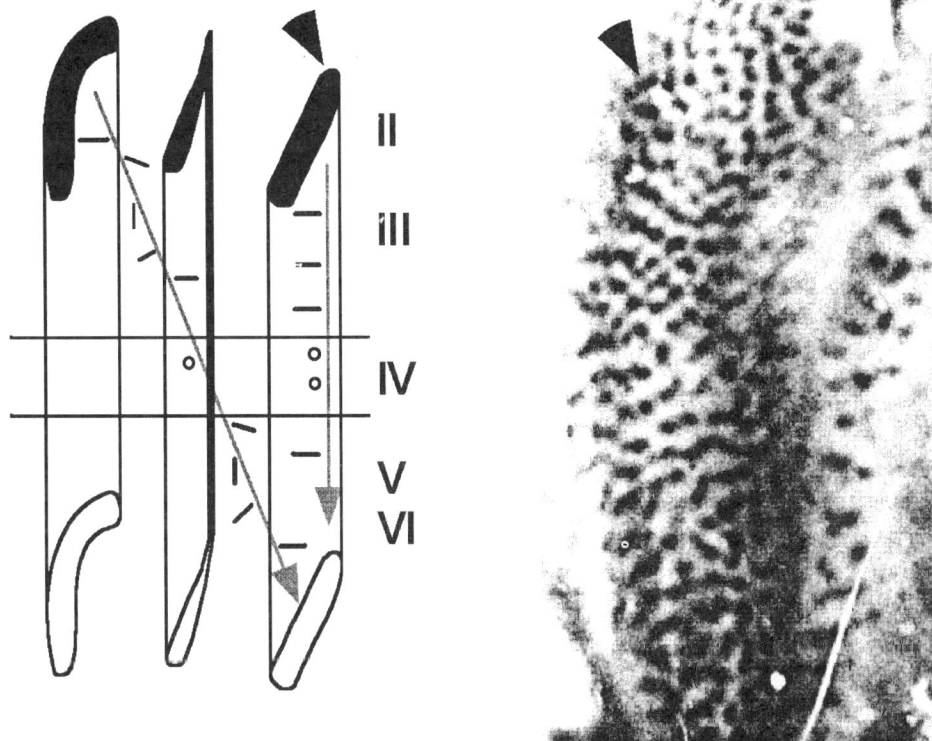

FIGURE 40.

(Left) Schematic reconstruction of two electrode tracts through area V1 in the macaque brain. The oblique tract encounters several iso-orientation columns (only three of which are shown as dark-surfaced solids, for simplicity). The orientation preference of neurons from which recordings were made is indicated by short lines along the path of the electrode. Notice the regular progression of stimulus orientation preferences as the electrode moves laterally through cortex. Within a given orientation column, the predominant orientation preference is found across all layers, but layer IV (as demonstrated by the vertical electrode path). Neurons in layer IV typically have predominantly concentric, unoriented receptive fields. The majority of cells in the three columns shown in the figure display preference (i.e., display maximum discharge) for horizontally oriented light slits. The surface distribution of columns that contain cells with similar orientation preferences, i.e., iso-orientation regions, is revealed using metabolic imaging with 2-deoxyglucose (Right). In this case, stimulation with horizontal lines leads to increased metabolic activity in cortical neurons with preference for horizontally oriented contours. These cells absorb larger quantities of the radioactive form of glucose. Then the brain is sliced and photographed. Given that previously active cells are grouped together, they appear as stained patches on the cortical surface. Several of these patches are present to form a complete representation of the contralateral visual field. Within each patch, there is also a regular alternation of eye preference.

This substance is absorbed in larger quantities by nerve cells which, by showing increased electrical activity, have increased metabolic demands. Cortical activation is induced by visual stimulation with vertical lines of a particular orientation for a short period of time. The cortex is then cut tangentially through non-afferent layers and placed onan x-ray film. The resulting photographs show patches of fluorescent cortex that presumably correspond to iso-orientation columns containing active cells. A new technique, optical imaging, can provide images of the surface distribution of orientation preference columns in the living brain of experimental animals. This method has significant advantages over traditional electrophysiological and autoradiographic techniques. For instance, as illustrated in Figure 41, it allows mapping of iso-orientation patches over a wide range of stimulus orientations in the same animal. In contrast, the 2-deoxyglucose technique can reveal iso-orientation patches for a single stimulus orientation (i.e., horizontal). Compared to electrophysiology, optical imaging can reveal the spatial layout of orientation-specific regions over a large portion of the cortical surface, as opposed to the limited area covered by a few electrode paths. When combined with electrophysiological recordings, optical imaging can be a powerful tool for the study of visual cortical function.

As previously mentioned, extensive horizontal connections have been found in area V1. The axons of many pyramidal cells may extend over a distance of 6 to 7 mm within the same cortical layer. Obviously, these projections interconnect cells with receptive fields that sample different portions of the visual field. It has been shown that these long axons form synaptic clusters at regular intervals with other cells in the same layer.

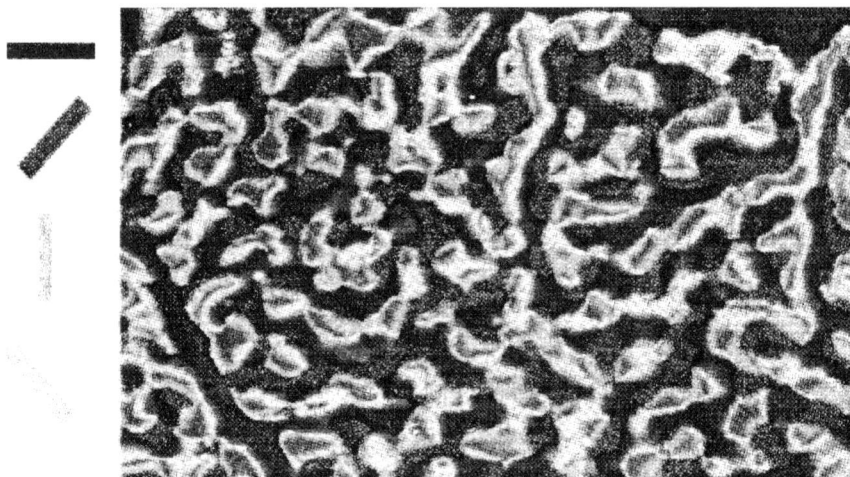

FIGURE 41.

Iso-orientation patches in the macaque visual cortex (area V1) revealed with optical imaging. The area shown measures 6 by 9 mm. The image is a composite of four photographs, each obtained during a 30 minute stimulation with visual contours at one of four different orientations (horizontal, vertical, and oblique). Cortical patches that showed greater activity during stimulation with a particular orientation have the same signal intensity (brightness) as the sample stimuli on the left-hand portion of the figure. (Reprinted with permission from Ts'o et al. (1990). Functional organization of primate visual cortex revealed by high resolution optical imaging, *Science, 249*, pp. 417-420. Copyright 1990 by the American Association for the Advancement of Science).

Histological studies using retrograde tracers, combined with metabolic autoradiographic techniques (utilizing 2-deoxyglucose), show that these pyramidal cells form connections with neurons that have similar orientation preferences (Gilbert & Wiesel, 1989, 1994). Electrophysiological studies corroborate these findings by demonstrating that cells in columns with similar orientation preferences have the tendency to fire in synchrony (Hata et al., 1991; Ts'o & Gilbert, 1988). The discharges of these cells, located a considerable distance from each other, are strongly intercorrelated, suggesting that the cells are interconnected.

These connections may be important for integrating single-unit responses produced by stimuli that extend beyond the receptive field of any individual cell. Thus, they may play a role in producing the *context effects* found in the orientation preferences of striate cells (Van Essen et al., 1991). This property has been explored by presenting simple stimuli, which consisted of a line segment oriented either vertically or horizontally within the classical receptive field of cells in area V1 in alert monkeys. Several line segments that were either parallel or horizontal to the central stimulus were simultaneously presented around it. A substantial proportion of the cells studied in this investigation showed marked response suppression when the surrounding element had the same orientation as the central stimulus. The discharge of these cells was not significantly affected when the orientation of the center stimulus and that of the surrounding line segments were orthogonal to each other.

Horizontal connections in V1 have also been implicated in processes of binocular interaction, two of which have received extensive attention: binocular facilitation and interocular suppression. The former is manifested by an increase in the firing rate of neurons in response to gratings of similar orientation presented to both eyes. The latter is defined as decreased firing when gratings of substantially different orientations are presented to each eye.

Although conclusive evidence regarding the role of horizontal connections between cells with similar orientation preferences in the brain mechanism(s) responsible for visual perception is still lacking, it is likely that they contribute to a variety of context-dependent phenomena (Gilbert & Wiesel, 1994). Examples of such phenomena are instances in which the perception of a particular sensory cue such as brightness, color, or orientation is influenced by the value of corresponding cues in their immediate vicinity (e.g., Butler & Westheimer, 1978). Further, it has been suggested that binocular interactions, documented at the neuronal level, play a crucial role in the brain mechanism(s) responsible for certain perceptual phenomena of binocular vision, like binocular fusion and stereopsis, and binocular rivalry. Electrophysiological evidence for such interactions has been obtained in the form of binocular facilitation (Pettigrew et al., 1968) and interocular suppression (Leopold & Logothetis, 1995).

Iso-orientation and ocular dominance columns are arranged in a systematic fashion with respect to each other. Thus, oblique or tangential microelectrode penetrations reveal a systematic alternation of orientation preference and ocular dominance. Within each ocular dominance column there is a more or less complete representation of orientation preferences. This representation is repeated in adjacent columns which contain cells dominated by the other eye (Hubel & Wiesel, 1974; Wiesel & Hubel, 1974). Hubel and Wiesel used the term *hypercolumn* to describe a "block of cortex" approximately 1 mm^2 that contains a complete representation of orientation preferences and a complete cycle of ocular dominance. Adjacent hypercolumns receive input from adjacent retinal loci. Thus, each hypercolumn is equipped to perform a complete analysis of a small portion of the visual field with respect to contour, color, direction of motion, eye of origin, and retinal disparity.

Sensitivity of neurons to visual contours

Cells in area V1 can be readily classified on the basis of the layout of their receptive field and their preference for simple visual stimuli (such as bars, slits, lines, and edges) to three main classes: *simple, complex,* and *hypercomplex.* This classification scheme was originally proposed by Hubel & Wiesel (e.g., 1968) and is still generally accepted to this day. The most characteristic properties of each cell class are summarized below.

Simple cells

Simple cells are found predominantly in layer IV. They have receptive fields with clearly separated excitatory and inhibitory zones. Typically they display an excitatory band (ON region) surrounded by two wider inhibitory regions (OFF regions). Other cells show a single ON- and a single OFF-region positioned side by side (see Figure 42). Preferred stimuli include light slits and visual edges of a particular orientation aligned with the excitatory portion of their receptive field.

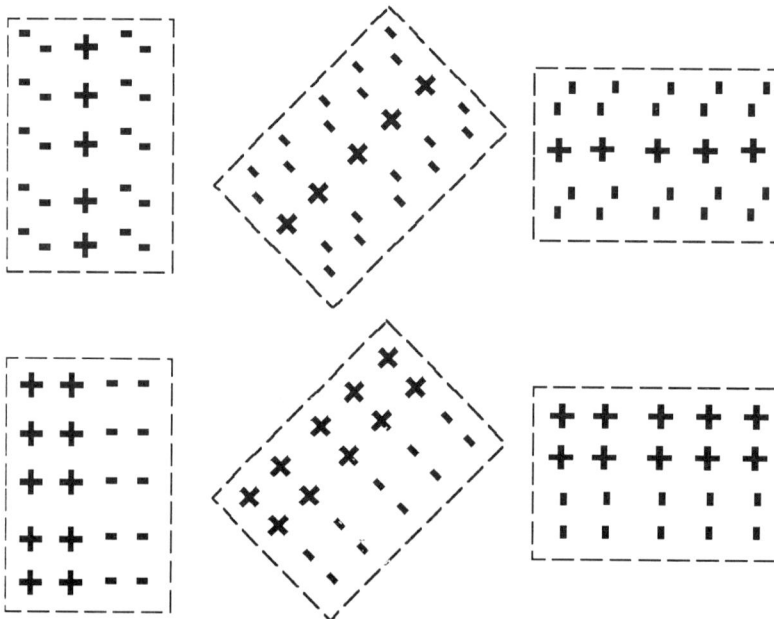

FIGURE 42.

Examples of receptive fields of simple cells. Regions which, when stimulated by light, are associated with increased firing rate are indicated with "+" signs (ON-regions). The opposite effect is induced when light falls on regions marked by "-" signs (OFF-regions).

Simple cells show phase sensitivity: they cease firing in response to a 90° phase shift of a luminance edge relative to the excitatory region of their receptive field (*null position*) and show maximal inhibition to a complete phase reversal (Figure 43, see also Figure 32). Another important feature of the responses of simple cells is that the degree of preference for a particular orientation (i.e., the cells' orientation tuning) is minimally affected by stimulus contrast (Sclar & Freeman, 1982). The size of the receptive field of a typical simple cell located at or near the foveal representation ranges between 0.5° and 1.0°. Most cells are driven by one eye only (Hubel & Wiesel, 1968). In sum, simple cells appear to be specialized for conveying information regarding the spatial orientation of a visual stimulus regardless of stimulus contrast.

Many simple cells show sensitivity to stimuli that contain wavelength transitions (known as *isoluminant wavelength contrasts*[6]) (e.g., Thorell et al., 1984). Tested with such stimuli, most cells can be classified as *double opponent*. For instance, some double opponent neurons are excited by green light falling on the ON-region and inhibited by red light on the ON-region. They display opposite responses when the OFF-region is stimulated. Interestingly, the organization of the receptive field into subregions is very similar for pure luminance (i.e., black and white) and for wavelength contrasts, suggesting that stimulus luminance and wavelength may be analyzed, at least partially, by overlapping populations of cells. Similar findings have been reported for dLGN cells. When tested with wavelength contrasts, double-opponent cells show phase sensitivity: they demonstrate null position effects with 90° phase shifts, as they do with pure luminance contrasts.

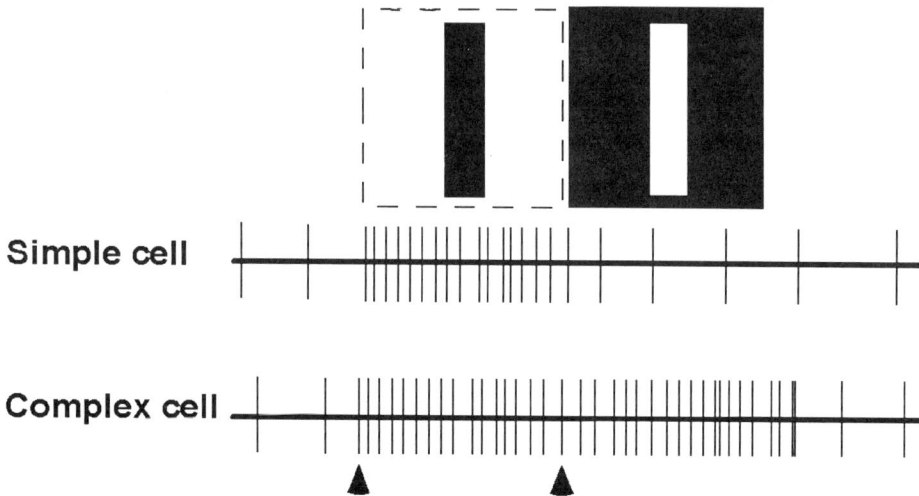

FIGURE 43.
Schematic representation of the response profile of a double-opponent simple cells in area V1 and of a complex cell. Each vertical line represents a single spike (extracellular recordings from single cell). Changes in firing rate are reflected in the density of vertical lines. Notice the sensitivity of the former to the phase-reversal of the stimulus, that is not evident in the response of the complex cell. Similar profiles are found with color stimuli. (Based on data from Thorell et al., 1984).

The remainder of the simple cells can be classified as *simple opponent*: they show excitatory responses to stimuli of a particular wavelength that cover a large portion of the receptive field (i.e., full-field stimuli) as well as obvious inhibitory responses to full-field stimuli reflecting the opponent wavelength.

Complex cells
Complex cells are exclusively encountered in the non-afferent layers II, III, V, and VI (Hubel & Wiesel, 1968; Michael, 1985). In contrast to simple cells, they respond to bars, slits, edges, or gratings of a particular orientation anywhere in their receptive field. Typically they respond better to slow moving rather than stationary stimuli, and many cells show preference for a narrow range of directions of visual motion. As illustrated in Figure 43, the majority of complex cells are not sensitive to the sign of a luminous bar or edge. The response of complex cells remains unaffected when a linear stimulus is extended beyond the excitatory region of their receptive field.

The majority of complex cells do not show the classical color opponency. A substantial proportion of these cells respond to a wide range of isoluminant wavelength contrasts, as well as to pure luminance contrasts. Some of these cells show preference for certain wavelengths over others (when tested with color bars) and display similar wavelength sensitivity anywhere in their receptive field. About one third of complex cells do not show preference for a particular wavelength, although they may respond equally well to isoluminant wavelength contrasts and to pure luminance contrasts (these units were named *universal color cells*, see Figure 44). Thus, although most complex cells apparently convey little or no information regarding stimulus hue, their responses could convey contour information, even at isoluminance.

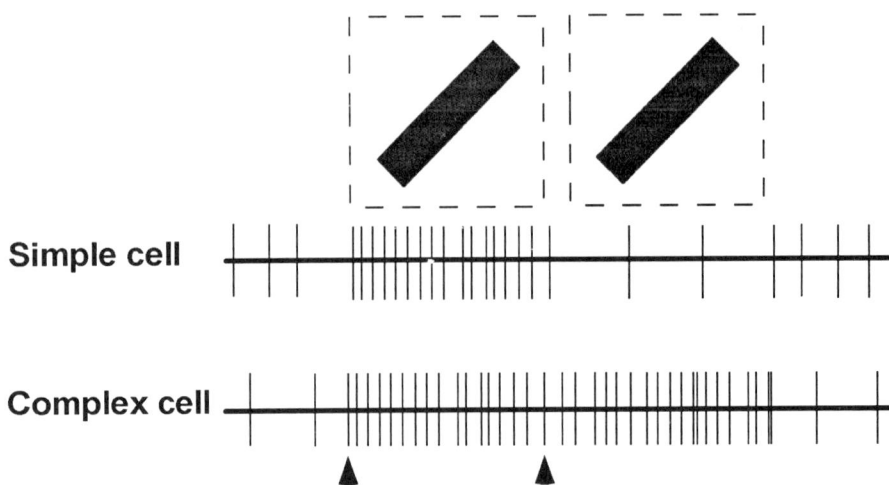

FIGURE 44.
Schematic representation of the typical response profile of a simple and a complex cell in area V1 to a red-green transition. (Based on data from Thorell et al., 1984).

End-stopped (Hypercomplex) cells
Like complex cells, end-stopped cells are found predominantly in the superficial layers. As their name implies, they differ from complex cells in that the excitatory region of their receptive field shows clear borders (Hubel & Wiesel, 1968; Hubel and Livingstone (1987). As shown in Figure 45, extending a line stimulus (a bar, slit, or edge) beyond that region produces response suppression, the magnitude of which depends upon the actual length of the stimulus.

End-stopped cells

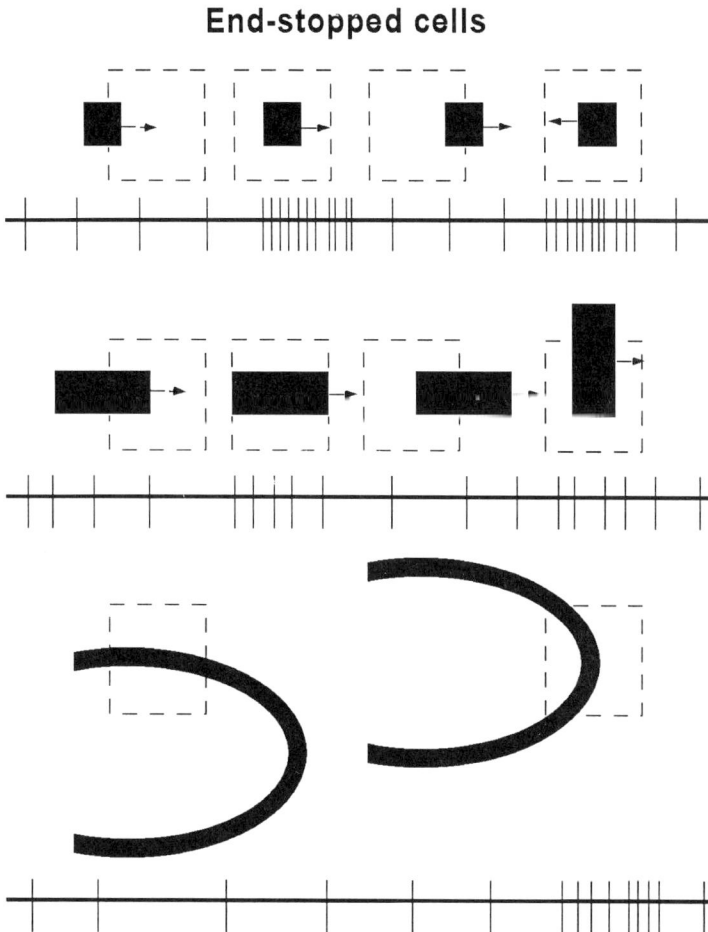

FIGURE 45.
Schematic representation of typical response profiles of an end-stopped cell in the macaque area V1. Each vertical line represents a single spike. This cell responds more vigorously to a small square positioned at the center of its receptive field (top row). Extending the surface area of the stimulus horizontally leads to a significant reduction in firing rate, which is further reduced when the stimulus crosses the receptive field boundary (middle row). The same neuron responds more vigorously to certain portions of a curved contour than to others. (Based on data from Hubel & Wiesel, 1968 and Hubel & Livingstone, 1987).

Thus, they are suitable for signaling curved contours (Hubel & Livingstone, 1987). Some estimates bring the proportion of end-stopped cells in Area V1 to 20%. In all other respects, the response properties of end-stopped cells closely resemble those of complex cells. In conclusion, whereas cells in the retinocortical pathway are primarily sensitive to local contrast transitions in the visual environment, neurons in V1 are capable of coding contours within a narrow range of stimulus orientation and size.

These findings have reinforced the idea that individual neural units in the visual cortex operate as *feature detectors*. According to this view, the role of simple cells is to signal the position of local luminance transitions (such as edges and bars of light) of a particular orientation. Complex cells appear suitable for signaling the orientation of small line segments that are part of a visual scene regardless of their location in the animal's visual field. Finally, hypercomplex cells are implicated in the detection of contours of a particular orientation and size. Implicit in this formulation is the notion of a hierarchy of connections between cells that belong in different functional classes. For instance, the spatial arrangement and orientation preferences of a group of simple cells would determine the receptive field characteristics of the complex cell to which they project. However, subsequent investigations have revealed a number of problems with this simple version of the "feature detector" hypothesis. Although complex cells receive a major input from simple cells, their firing does not depend exclusively upon those connections (Hammond, 1985). Further, experiments using moving textured arrays in isolation or in combination with traditional edge or bar stimuli indicate that the firing of a single complex cell cannot provide an unambiguous signal of the presence of these stimuli (Hammond & MacKay, 1975).

An alternative to the "feature detector" model is the *spatial frequency coding* theory, according to which, visual cells operate as spatial frequency filters. Changes in the firing of simple and complex cells signals the presence of specific Fourier components[7] in a complex visual pattern. At later stages, the responses of many cells that display a wide range of spatial frequency preferences are integrated in order to reconstruct the entire visual stimulus. Although attractive in its ability to account for the detection of complex visual patterns regardless of their location within the visual field in mathematical terms, unfortunately there is no conclusive evidence that the visual system utilizes spatial frequency information to support perception.

Many neurons in area V1 are sensitive to retinal disparity cues. A description of the response properties of these cells has been incorporated in the section on area V2 that follows.

Direction selectivity
A prominent feature of many neurons in the visual system is their ability to vary their firing rate in response to a change in the direction of motion of a visual stimulus. This property is called *direction selectivity*. Estimates of the proportion of direction-selective cells in area V1 range between 20% and 30% across studies (De Valois et al., 1982; Hawken et al., 1988; Snowden et al., 1991). Hawken and colleagues (1988) found highly selective cells in the upper subdivisions of layer IV (i.e., sublaminae IVA, IVB, and IVCa) and in layer VI. The highest concentration of direction selective cells was found in lamina IVB.

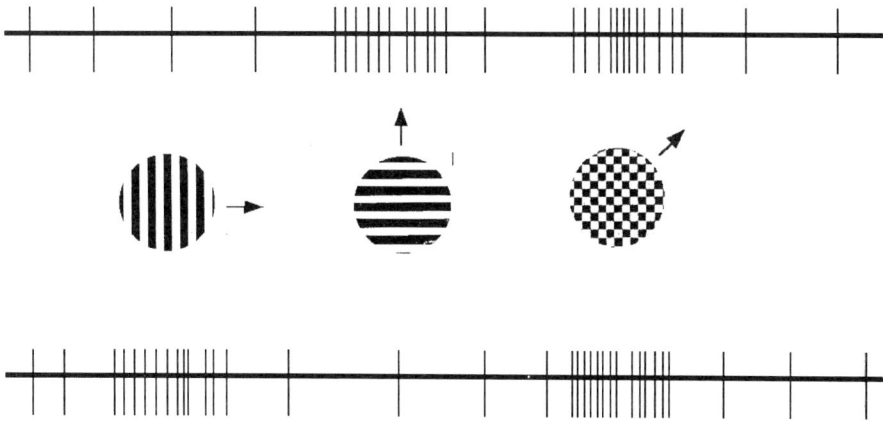

FIGURE 46.

Responses of two V1 cells to each of the two components (left and middle patterns) and to the plaid composite (left). The arrows indicate the perceived direction of motion. The cell on top displays clear preference for upward motion and the cell at the bottom shows preference for rightward motion. Neither cells' discharge patterns are affected by superimposition of the two gratings, although the perceived motion of the plaid is not in the preferred direction for either cell. (Based on data from Movshon et al., 1985).

The bulk of the projections from layer IVB to the middle temporal area (a region specialized for analyzing certain cues for visual motion that is discussed in greater detail later in this chapter), originate within clusters of cells that show direction selectivity.

Typically, V1 cells respond to individual components of complex moving arrays. Two examples clearly demonstrate this property. In one study (Movshon et al., 1985), recordings from V1 cells were made in two conditions. In the first, the stimulus was a single sinusoid grating moving in one direction. In the other condition, two gratings, with bars orthogonal to each other and each moving in a direction perpendicular to the direction of motion of the other, were superimposed forming plaid patterns that appeared to be moving in yet a third direction, as shown in Figure 46. It was found that V1 cells responded more vigorously to the direction of motion of the individual component gratings rather than the perceived direction of the plaid pattern. Integration of the responses of neurons that show selectivity for component motion direction occurs at a latter stage along the visual pathway. Another form of complex motion that has attracted the attention of researchers is *transparent motion* (e.g., Snowden et al., 1991). The latter is present in many natural scenes. One example is the view of the surroundings through the window of a moving car, while water streaks down the glass. Transparent motion poses a special problem because it requires that the visual system must be able to respond to each motion component (or *motion vector*) separately and, at the same time, produce a unified percept. To simulate transparent motion, Snowden and colleagues used stimuli consisting of two spatially overlapping dot arrays that moved in opposite directions (Snowden et al., 1991). Human observers perceive these stimuli as consisting of two sliding surfaces (see Figure 47). The discharge rate of the majority of cells in area V1 was not affected by the presence of motion in a direction opposite to the preferred direction. In other words, V1

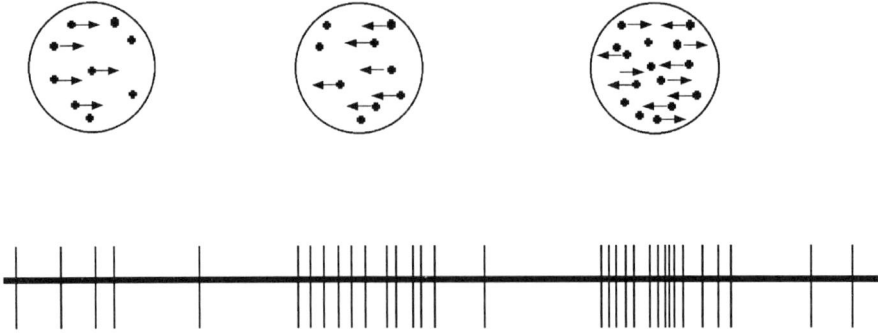

FIGURE 47.
Typical responses of a V1 cell to an array of moving dots. In the middle and left-hand displays all dots move in one direction. The display on the right is a composite of the two single-motion arrays. Notice that the cell shown here has a clear preference for the leftward motion display, but its firing rate is not reduced when motion in the opposite (or antipreferred) direction is added in the composite display. (Based on data from Snowden et al., 1991).

Simple cell

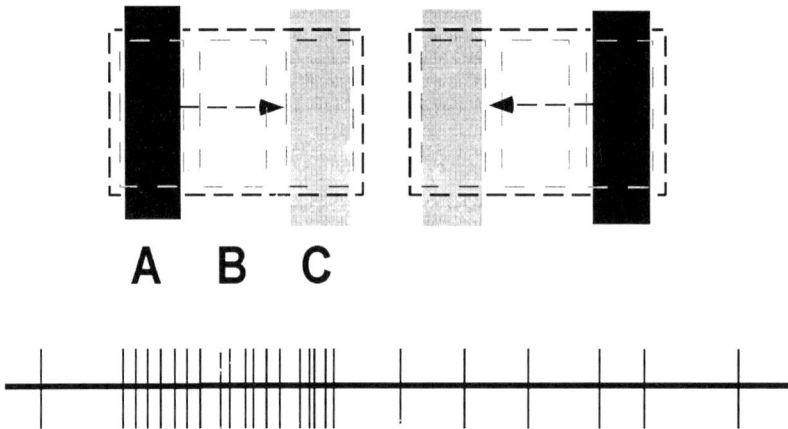

A   B   C

FIGURE 48.
Schematic representation of receptive field organization that could account for the direction selectivity of primate V1 cells. A moving stimulus (such as vertical bar) stimulates the different receptor field subunits in a particular order, for instance subunit A followed by subunit B, and then subunit C. In this cell, which shows preference for rightward motion, stimulation of subunit A triggers an excitatory response that further enhances the cell's response to subunit B, provided that A is stimulated before B. The same applies for the ordered stimulation of subunits B and C. In that case, visual motion from A to B to C is the cell's preferred motion direction. Conversely, no response enhancement occurs when the visual stimulus moves from C to B to A, that is, in the cell's antipreferred direction.

neurons appear capable of signaling the presence of each motion vector separately at a particular locus in the visual field.

In general, motion selectivity in cortical cells is believed to reflect the outcome of the integration of several non-selective neural inputs. The latter are spatially organized to form functional subregions within the cell's receptive field, like the ones shown in Figure 48. Integration of inputs from the different subregions may involve facilitatory or suppressive processes, or both. Thus, inputs from subregion A may facilitate an excitatory response to simulation of subregion B, only if the input from A arrives on the cell before the input from B. In the example of Figure 48, this would occur when a stimulus is moving from left to right. When the stimulus is moving in the opposite direction, and stimulation of subregion B precedes subregion A stimulation, neither input is sufficient to elicit a response from the cell. *Response suppression*, on the other hand, implies an inhibitory process, whereby stimulation of subregion B either, altogether prevents, or simply counteracts the excitation induced by stimulation of subregion A. Physiologically, the former process would correspond to the linear addition of EPSPs and IPSPs induced in a given cell by the various subregion-specific inputs. The latter process amounts to shunting inhibition. In this case, EPSPs generated at the dendrites is prevented from reaching the axon hillock, by synaptically induced increase in membrane conductivity. As in the case of orientation selectivity discussed below, the inhibitory, or sign inversing input, is probably mediated by cortical interneurons.

Although inhibitory processes are very likely involved in shaping direction selectivity in cat primary cortex cells (Creutzfeldt et al., 1974; Sillito, 1975), in primates, a mechanism of direction selectivity based on response suppression fails to account for the typical response of V1 cells to transparent motion stimuli. Response suppression models predict that a visual display that contains elements moving in the anti-preferred direction will lead to response suppression. This behavior is rarely observed in V1 cells (Snowden et al., 1991), a finding that is consistent with a facilitatory process. Inhibitory processes are likely to be involved in determining the motion-sensitive responses of neurons in association visual cortices, like area MT, which will be discussed in a subsequent section.

Orientation selectivity

As previously mentioned, one of the most striking characteristics of the primary visual area is its columnar organization with respect to orientation sensitivity. As a recording electrode moves laterally through the cortex above or below the afferent lamina (IV), the stimulus orientation that produces the largest response (i.e., the cell's preferred orientation) changes systematically (Hubel & Wiesel, 1962; 1968; see Figure 40). Often several cycles of preferred orientation are encountered in the course of a single electrode track. Typically, cells increase their firing rate in response to a range of stimulus orientations and they often show response suppression to contours oriented perpendicular to the preferred orientation (i.e., the cell's anti-preferred orientation). Cells with similar orientation preferences tend to cluster together in patches or iso-orientation columns. In primates, cortical domains characterized by similar orientation preferences are not continuous: they are interrupted by regions containing cells with poor orientation sensitivity. As previously mentioned, iso-orientation domains overlap to a great extent with cortical patches that are heavily stained for cytochrome oxidase (i.e., blobs). Conversely, cortical patches that show poor orientation sensitivity contain very small amounts of this enzyme (Hubel & Livingstone, 1987).

A variety of proposals have been put forward in order to outline the neuronal mechanism responsible for the emergence of orientation selectivity in the non-afferent layers of area V1. Excitatory connections between geniculocortical afferents and their lamina IV target neurons are part of every proposal. In addition, intrinsic interconnections among cortical neurons play a crucial role in certain models. Models differ, however, in the degree and type of inhibitory intracortical connections that they postulate (for a detailed review, see Ferster & Miller, 2000). The earliest attempt Hubel and Wiesel (1962), to account for this property invoked predominantly excitatory synaptic connections. According to this hypothesis, the receptive fields of the afferent fibers that converge onto a single cortical cell are arranged to form a straight line (see Figure 49). A light slit that falls onto all of the individual receptive fields induces maximal excitation. Any other stimulus orientation would excite fewer retinal ganglion neurons and reduce the excitatory drive on the simple cell.

This model became the standard account of orientation selectivity in most texts published in the last three decades (see Kandel, 1991). One of the model's strongest assets is its inherent simplicity. Furthermore, it is compatible with histological data regarding the distribution of geniculocortical projections. Convergence of ascending fibers onto a single cortical neuron appears to be the rule in striate cortex. Moreover, there is ample evidence for the existence, at least in the cat, of direct connections between geniculocortical afferents and simple cells (Ferster & Lindström, 1983). The spatial distribution of these contacts within the dendritic field of recipient cells is also consistent with the model outlined in Figure 49 (Reid & Alonso, 1995). Finally, the orientation tuning of simple cells is not significantly affected by silencing cortical activity, suggesting that geniculocortical input alone is sufficient to induce orientation selectivity (Chung & Ferster, 1998).

A serious limitation of any model that relies exclusively on excitatory connections is that it cannot account for the relative invariance of the responses of simple cells across a wide range of stimulus contrast levels. A cortical neuron that receives solely excitatory afferent input would become excited to the same degree by a high-contrast slit in the non-preferred direction as it would by a low-contrast slit in the optimal orientation (Martin, 1988). In reality, however, the orientation tuning of simple cells do not vary significantly with stimulus contrast (Skottum et al., 1987). This property can be accounted for by introducing an inhibitory component into the circuit proposed by the excitatory model. Inhibitory input provided through intracortical connections onto simple cells was first proposed by Hubel and Wiesel (1962) to explain the discharge suppression (i.e., a reduction in firing rate below baseline levels) often observed in response to stimuli in the cell's antipreferred orientation. Intracellular recording data were instrumental in supporting this view by permitting researchers to determine the independent contribution of excitatory (in the form of EPSPs) and inhibitory (in the form of IPSPs) inputs that converge onto a given simple cell. Using this technique, it was found that IPSPs in many simple neurons are indeed tuned to the cell's preferred orientation (Ferster, 1986). This suggests that simple cells receive inhibitory input from other cortical neurons that show similar orientation preference. More recent studies have revealed that the receptive fields of the inhibitory cortical neurons are aligned and overlap spatially with the receptive field of the simple cell.

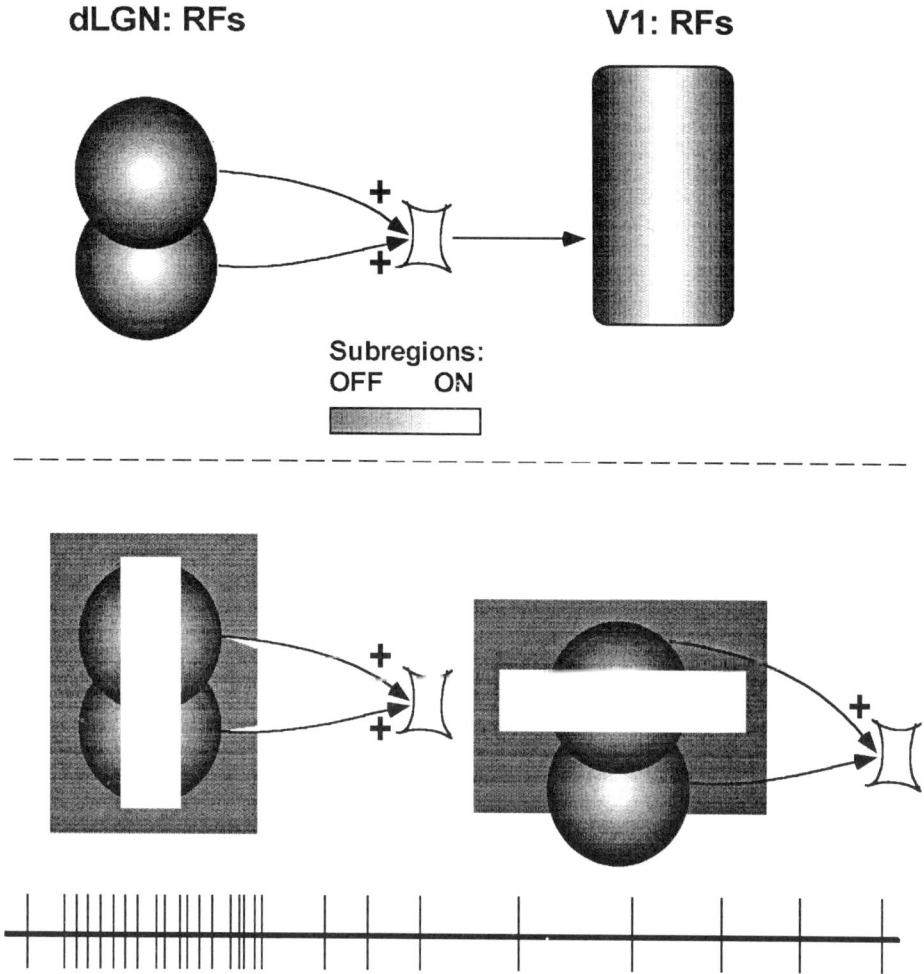

dLGN: RFs

V1: RFs

Subregions:
OFF   ON

FIGURE 49.

(*Top*) Schematic drawing of the layout of the concentric receptive fields of geniculocortical afferents that project onto a simple cell in area V1 according to the excitatory model of orientation selectivity. Excitatory (ON-) subregions are indicated by "+" signs while OFF-regions are indicated
by "-" signs. (*Bottom*) Hypothetical discharge patterns of the same simple cell. The stimulus (a slit
of light) is superimposed on the receptive fields of the afferent geniculocortical neurons. The cell
responds vigorously to the vertical slit because it stimulates the ON region of all three dLGN neurons, which in turn provide synchronous excitatory input to the simple cell. The simple cell's response drops significantly when the stimulus is turned by 90° because the cell now receives excitatory input from only one afferent. In fact, many simple cells in V1 display strong response suppression with stimulation in the antipreferred direction. This effect can be seen clearly using intracellular recordings (not shown in this figure).

dLGN: RFs   V1: RFs

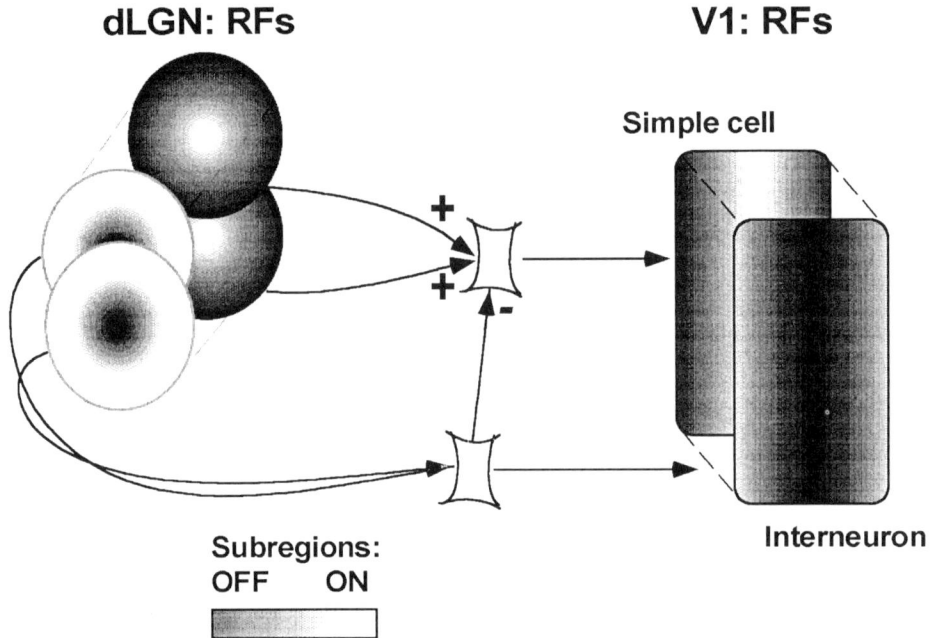

FIGURE 50.
Schematic representation of the excitatory/inhibitory model of orientation selectivity. The simple cell shown here receives direct geniculocortical input from an array of dLGN cells as specified by the excitatory model shown in Figure 49. It also receives strong inhibitory input from cortical interneurons (only one shown here), the receptive field of which is opposite in spatial phase. (Based on data from Hirsch et al., 1998 and Troyer et al., 1998).

As illustrated in Figure 50, two receptive fields have opposite polarity with respect to the spatial layout of the ON and OFF subregions: the ON subregion of the simple cell overlaps spatially with the OFF subregion of the inhibitory neuron, whereas the OFF subregion of the simple cell is aligned with the ON subregion of the inhibitory cells.
Inhibitory interneurons are abundant in the cortex, the most prominent type in area V1 being smooth stellate cells (Freund et al., 1983). One type of smooth stellate cells, in particular, basket cells, are the most likely candidates for providing orientation-specific inhibition. These cells are encountered in all layers forming synaptic contacts with the somata and the dendritic spines of spiny stellate cells. It has, however, been difficult so far to examine the receptive field characteristics of V1 interneurons in sufficient detail to see if they are fully consistent with the requirements of the model of orientation selectivity described above. The model outlined in Figure 50 is consistent with data regarding the orientation tuning of simple cells and can also account for the capacity of these cells to maintain accurate coding of stimulus orientation over a wide range of stimulus contrast levels. Given the scarcity of detailed examinations of the electrical behavior of simple cells in the primate area V1, proposals regarding orientation selectivity were modeled after the visual cortex of the cat.

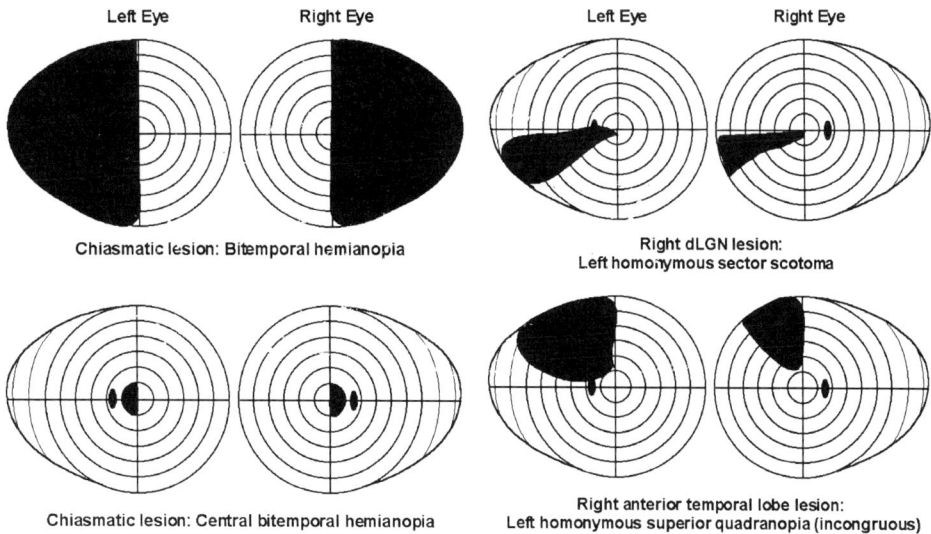

FIGURE 51.
Types of visual field defects arising from lesions that affect the retinocortical pathway. The dark areas indicate locations where pattern vision is significantly impaired. *(Top left set of images)* The most common form of visual defect following a chiasmatic lesion, like one caused by a tumor, involves a significant portion of the temporal hemifield bilaterally. The damaged fibers originate from the nasal hemiretinas. When the lesion affects only nasal fibers that originate from the foveal region, the visual deficit is restricted to the temporal portion of the central visual fields *(bottom left set of images)*. Lesions to the dLGN are relatively rare and often produce incongruous (but always homonymous) visual field defects (i.e., the defect occupies a different portion of the visual hemifield in each eye *(top right set of images)*. Similar defect can be seen following damage to the optic tract. Anterior temporal lobe lesions often disrupt optic radiation fibers (Meyer's loop) that originate in the lower part of the retina, causing deficits in the upper visual field quadrants contralateral to the lesion *(bottom right set of images)*.

An important difference between the primate and cat visual cortex is the presence of oriented, simple cells in layer IV in the latter species. In primates, cells that show comparable orientation selectivity are only found in non-afferent layers.

Contributions of the primary visual cortex to vision
Lesions at different points along the visual pathway produce predictable visual deficits (see Figures 51-52). The importance of the calcarine cortex for vision in humans is best demonstrated by the dramatic effects of damage to the occipital pole. Depending upon their size and location these lesions can cause partial or even a complete loss of pattern vision in a portion of the visual field. Unilateral lesions, encompassing both the ventral and the dorsal bank of the calcarine fissure, cause *homonymous hemianopia*[8] in the contralateral visual field. Partial lesions that spare a portion of the primary area may cause *homonymous quadranopia*. A unilateral lesion in the ventral bank of the calcarine sulcus affects vision in the upper quadrant of the contralateral visual field. Damage to the visual cortex in the dorsal bank of the calcarine fissure leads to visual deficits in the lower

quadrant of the contralateral visual field. Smaller lesions can cause loss of vision restricted to a smaller portion of the visual field (a *homonymous scotoma*). Bilateral lesions of the optic radiations or area V1 typically lead to loss of conscious visual sensation.

Such lesions are more commonly caused by an infarction in the basilar artery or by bilateral occlusion of the posterior cerebral artery, but also by occipital tumors, and subarachnoid hemorrhage.

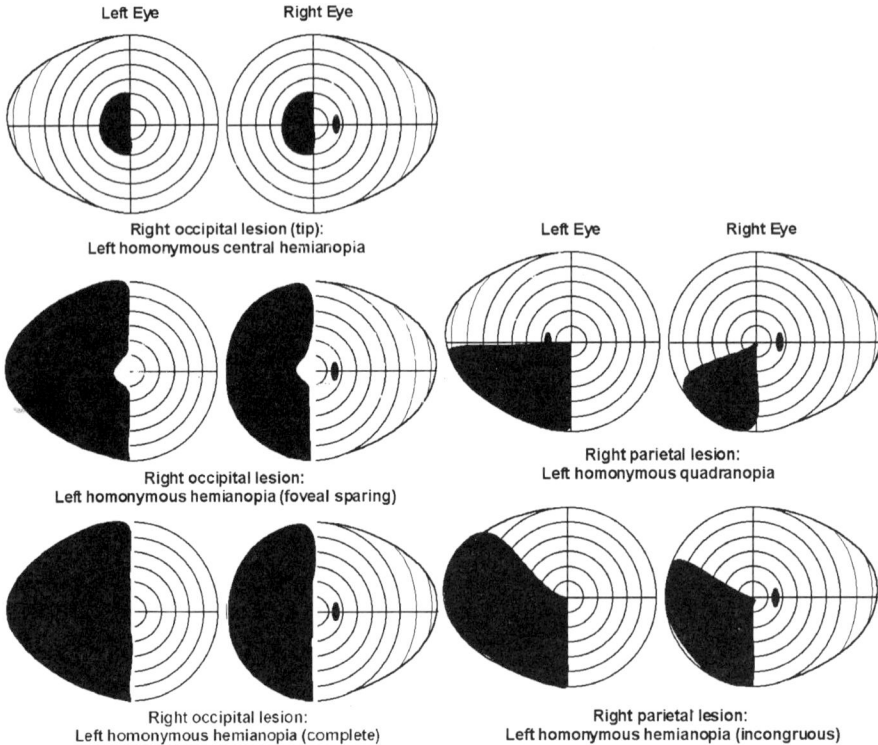

FIGURE 52.

(Left column) Visual field defects arising from unilateral lesions in the occipital lobe. When the lesion is restricted to the tip of the occipital lobe that contains the foveal representation, the defect may be limited to the central portion of the (contralateral) hemifield *(top)*. A homonymous hemianopia that spares the foveal representation can occur when (a) the tip of the temporal lobe is not affected by the lesion, or (b) in a stroke that only involves the posterior cerebral artery. In the latter case, blood supply to the portion of the primary visual cortex that contains the foveal representation can be maintained through terminal branches of the middle cerebral artery *(middle)*. When the lesion encompasses the entire primary visual cortex the hemianopia is complete *(bottom)*. In all cases a congruous hemianopia in the hemifield contralateral to the lesion is observed. *(Right column)* Parietal lobe lesions often disrupt optic radiation fibers that originate in the superior portion of the retina, and cause visual impairment primarily in the inferior quadrants contralateral to the lesion.

A number of anatomical and physiological changes in various other brain areas have been reported following damage to the primary visual area. The striate cortex is the major source of visual afferent input to the extrastriate visual areas. Therefore, it is expected that damage to, or even transient inactivation of, the primary visual cortex affects the visual responsiveness of neurons in these areas. Thus, cooling area V1 in primates causes a marked reduction of the ability of cells in areas V2 and V4 to respond to visual stimulation (Girard & Bullier, 1989; Girard et al., 1991). In contrast, visual responsiveness is largely preserved in two other extrastriate areas, the inferior temporal cortex and the middle temporal area. Closer examination revealed that the activity in the inferior temporal cortex depends upon visual input relayed from the contralateral striate cortex via the corpus callosum (Rocha-Miranda et al., 1975). It has also been found that preservation of the sensitivity of neurons in the middle temporal area to visual motion depended upon input from the superior colliculus (Rodman et al., 1989).

An interesting feature of visual field defects caused by lesions in area 17 (V1) in humans is that some patients are capable of responding to simple attributes of stimuli presented within a scotoma, although they report having no visual experience of these stimuli. This condition was termed *blindsight* by Weiskrantz and colleagues to indicate the fact that certain visual functions may be preserved in the absence of conscious awareness of the visual stimuli (Weiskrantz et al., 1974). A number of techniques have been employed to test the residual visual function in patients with blindsight. Typically, the patient is asked to indicate whether a stimulus has been presented or not (signal detection tasks), to point to the spatial location at which a visual stimulus was presented, or to choose the stimulus that was flashed within his "blind" visual field among a set of items presented in the intact portion of the visual field (Stoerig & Cowey, 1990). In a classic study, Weiskrantz and his colleagues (1974) examined a patient (DB) who suffered from a dense left homonymous hemianopia produced by an arteriovenous malformation in the right occipital pole. The patient could make accurate finger movements to reach the location of a light spot flashed within his "blind" field. In addition, he could discriminate a bright circle from a square presented within the scotoma with a fair amount of accuracy. However he was unable to make even the simplest perceptual discriminations when faced with stimuli that did not contain such simple sensory cues as orientation and local luminance transitions. For instance, he could not discriminate a square from a rectangle. In subsequent studies, patients with dense visual field defects were found capable of localizing visual targets by either pointing or by executing fast eye movements (saccades).

There is evidence that at least some of the residual visual functions in blindsight are subserved by extrastriate areas in the hemisphere ipsilateral to the lesion. This conclusion arises from the results of studies that compared the visual capabilities in two groups of blindsight patients. Patients in one group had unilateral lesions restricted to the primary visual area, whereas patients in the second group had undergone a complete removal of the neocortex in one hemisphere (hemispherectomized patients). Patients in the two groups showed similar performance in a variety of tasks. Typically, however, hemispherectomized patients had a severe deficit in detecting the direction of visual motion, a capacity that was preserved in many patients in the other group (Ptito et al., 1987). This finding suggests that at least one residual function, namely the ability to detect the direction of visual motion, may depend upon spared extrastriate areas in the ipsilateral hemisphere.

The finding that blindsight patients retain the ability to produce relatively accurate saccadic movements toward visual targets located in the periphery suggests that at least some of the residual visual functions in blindsight are mediated by the retinotectal system. This notion is consistent with the fact that residual visual functions in these patients can best be demonstrated with fast moving or flickering stimuli even at low contrasts. Under these conditions, visual input is transmitted mainly by the magnocellular pathway, which serves as a major source of visual input to the superior colliculus. The phenomenon of blindsight is therefore consistent with the proposal that the primate visual system consists of two subsystems that operate independently from each other. One system, the geniculostriate, appears to be primarily responsible for high resolution vision, which can trigger visual awareness. The primary visual cortex is an indispensable component of this system. The second system conveys sufficient information to permit accurate stimulus localization that could, in turn, support manual pointing and oculomotor orienting. Although very little empirical data exists in humans, this system probably involves the superior colliculus and its projections to the part of the thalamus known as the *pulvinar*. The latter is known to project to several visual association areas.

Visual association cortex: Area V2

In the macaque, area V2 is an elongated band of cortex located on the lateral surface of each hemisphere, forming the rostral border of area V1 (see Figure 21). V2 has been found in virtually all the primate species that have been studied anatomically (Kaas, 1993). In the human brain, area V2 corresponds to Brodmann's area 18. A large portion of this area lies on the medial aspect of the occipital pole and extends caudally into the postero-lateral surface of the hemisphere as it wraps around area 17. In primates, staining with cytochrome oxidase produces a distinct pattern of alternating thick and thin stripes separated by pale (less densely stained) interstripes (Hubel & Livingstone, 1987). Similar staining profiles have been found in fixed human brain preparations (Burkhalter & Bernardo, 1989). More on the functional significance of this finding is presented below. The functional characteristics of individual neurons in area V2 have been studied extensively in the macaque and the squirrel monkey by Hubel & Livingstone (1987). Four major classes of cells were identified in area V2: *simple unoriented, complex unoriented, complex oriented*, and *end-stopped* cells. These functional cell types were characterized by integrating information regarding their preference to multiple sensory cues.

Unoriented cells often show similar properties with unoriented cells in area V1. They are classified into three subtypes. *Broadband* cells display surround antagonism with the center of their receptive field, being either excitatory (ON-center cells) or inhibitory (OFF-center cells). Similar receptive field organization is noted when using isoluminant wavelength displays as well as black and white stimuli. In other words, their receptive field does not show opponent wavelength characteristics (i.e., excitation by a particular color and suppression of firing by a different wavelength). *Type II* cells have a wavelength-opponent center but lack an antagonistic surround. The following types of receptive fields are commonly encountered: Red-ON/Green-OFF (i.e., a red spot in the center of the receptive field excites the cell and a green spot suppresses its firing), Red-OFF/Green-ON, and Blue-ON/Yellow-OFF, Blue-OFF/Yellow-ON. Due to the lack of an antagonistic surround, a large spot of light is equally effective in driving these cells as a smaller spot of light that is constrained within the receptive field center. Finally, *double opponent* cells show wavelength opponent centers and antagonistic surrounds. Due to the presence of an antagonistic surround region, a large spot of light that covers the center as well as part of the surround area elicits weaker responses than a smaller spot that falls exclusively within the center of the receptive field.

The remaining unoriented cells can be classified as *complex unoriented*. Similar to the complex cells in area V1, these cells show the same wavelength sensitivity properties anywhere in their receptive field. Two subtypes have been identified: *broadband* and *color opponent* cells. The latter respond to spots of a particular wavelength and size, giving little or no response to larger spots. However, a small spot can excite the cell anywhere within its receptive field (which often can be as large as 4° of visual angle in diameter). When the center of the receptive field is stimulated simultaneously by two spots of light that contain opponent wavelengths (for instance a red and a green spot, for a Red-ON/Green-OFF cell), the firing rate of the cell does not change (mutual cancellation). Hubel & Livingstone (1985, 1987) reported a tendency for both types of unoriented cells to cluster in thin cytochrome oxidase-rich stripes.

The third major cell class in area V2 are *complex oriented cells,* which have very similar response properties with the complex oriented neurons in area V1. They respond best to bars and edges, demonstrating preference for a narrow range of stimulus orientations. They are not wavelength selective and their discharge is not affected by the sign of the luminance contrast (see Figure 49). Moving stimuli generally elicit a larger response than stationary stimuli, with a proportion of these cells showing selectivity for the direction of stimulus motion. Extending a bar or edge stimulus in either direction beyond the excitatory portion of the receptive field does not affect the strength of the neuron's response.

Finally, *end-stopped cells* correspond to the hypercomplex cells originally described by Hubel & Wiesel (1968) in the primate area V1. They typically show a near-square "activating region" and one or more inhibitory regions. Extending a linear stimulus beyond the excitatory region suppresses their response. Stimulation of the inhibitory region alone elicits no response on its own. These cells do not display clear color selectivity. In the squirrel monkey, complex cells tend to cluster together with end-stopped cells in regions that were lightly stained with cytochrome oxidase (pale stripes) and with disparity-sensitive cells in the thick cytochrome oxidase-rich stripes (Hubel & Livingstone, 1987).

Early work using retrograde staining with horseradish peroxidase injected in different V2 compartments suggested that thin stripes receive a major input from CO blobs on V1 (see for example, Hubel & Livingstone, 1987). Conversely, it was suggested that the majority of V1 inputs to interstripes and thick stripes originated in interblobs and layer IVB, respectively. On the basis of electrophysiological data, it was also suggested that V2 thin stripes contained neurons specialized primarily for the analysis of color, whereas neurons in V2 interstripes were primarily specialized for the analysis of luminance contours (form). More recent investigations, however, have revealed that thin stripes contain similar proportions of color-selective and orientation selective neurons (DeYoe & Van Essen, 1985; Levitt et al., 1994). Moreover, there are many indications that thin stripes may actually be organized into rather discrete regions that contain neurons selective for either chromatic or luminance contours (Roe & Ts'o, 1995). It remains to be seen if different subcompartments within V2 thin stripes receive segregated input from functionally distinct clusters of cells in V1 blobs. Recent studies using retrograde labeling have revealed a rather complex pattern of projections from V2 to areas V4 and MT, summarized in schematic form in Figure 53. Thus, it appears that different cell clusters within V2 thin stripes may give rise to segregated inputs that target distinct regions within V4. There is also evidence that inputs arising in V2 thin stripes and interstripes converge in the same V4 region (Felleman et al., 1997). Moreover, there is considerable overlap between the thick cytochrome oxidase-stained stripes in area V2 and those regions that contain neurons that become retrogradely labeled after horseradish peroxidase injections into area MT (Shipp & Zeki, 1985). These findings provide some evidence that the bulk of the projections from area V2 to area MT originate in thick stripes, whereas afferent projections from area V2 to area V4 originate in both thin stripes and interstripes.

Sensitivity to retinal disparity

Consistent with psychophysical evidence revealing a variety of phenomena that require binocular integration of visual input, electrophysiological studies have shown that the vast majority of striate and extrastriate neurons in the cat and monkey visual cortex receive input from both eyes. As previously described, inputs from the two eyes reach the primary visual cortex through separate routes. This segregation is maintained up to layer IV of the primary visual cortex. Subsequently, the two sets of inputs converge in the non-afferent layers of area V1.

Cortical binocularity is a prerequisite for the perception of stereoscopic depth cues. According to one estimate (Pettigrew et al., 1968), approximately 60% of the binocular neurons in the cat striate cortex are sensitive to retinal disparities present in linear contours (i.e. under conditions associated with local stereopsis). In primates, a similar proportion of binocular neurons respond selectively to various forms of disparity (Poggio & Fischer, 1977). Binocular cells have identical receptive field properties in the two eyes (Hubel & Wiesel, 1962). They become maximally excited when the stimuli presented to each eye are identical. However, the monocular receptive fields in a significant proportion of these neurons do not overlap completely. This property, that was termed "receptive field disparity"[9] (Nikara et al., 1968), can account for the sensitivity to local stereoscopic cues. Subsequent studies employing dynamic random dot stereograms have shown that a significant proportion of neurons sensitive to local stereopsis cues are also sensitive to purely binocular disparity cues, which are essential for global stereopsis. Disparity-tuned cells in the macaque monkey can be classified into four subtypes (Hubel & Livingstone, 1987; Poggio & Fischer, 1977; Poggio et al., 1985). *Tuned excitatory* cells are maximally excited when the horizontal retinal disparity between the images reaching the two eyes is very small in magnitude. The preferred magnitude of horizontal disparity varies across cells, with some neurons being sensitive to disparities as small as 3 min. of arc. Thus, tuned excitatory cells demonstrate maximal excitation in response to stimuli that are perceived as fused. *Tuned inhibitory* units show maximal inhibition in response to zero or near-zero horizontal disparities and maximal excitation at larger disparities. *Near neurons* show an asymmetric response profile: they are excited by crossed[10] disparities and inhibited by uncrossed disparities.

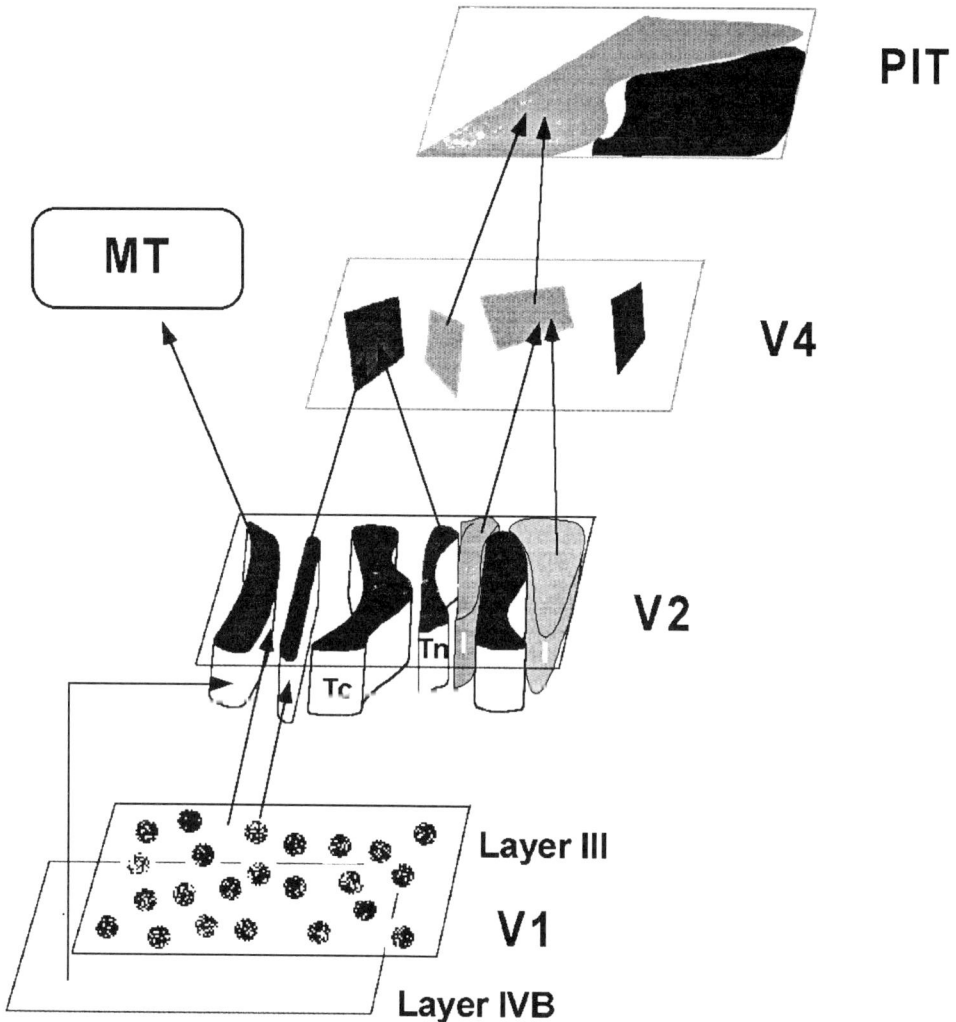

FIGURE 53.
Schematic representation of forward connections from area V1. Round patches in the lower figure represent lamina III blobs. Area V2 contains thick (Tc) and thin (Tn) heavily stained stripes and lightly stained interstripes (I). Based on the origin of the input they receive from V2, one can distinguish distinct patches within V4, displayed here as darker or lighter shapes. Distinct subregions have also been found within the posterior inferior temporal area (PIT) on the basis of the origin of afferent input from area V4. The projection from thick V2 stripes to the middle temporal area (MT) is also shown. (Based on data from Felleman et al., 1997).

Finally, *far neurons* show the opposite pattern of response (i.e., excitation with uncrossed and suppression with crossed disparities). The majority of neurons that show sensitivity to global stereoscopic cues are complex neurons (Poggio et al., 1985). The binocular cells that are exclusively sensitive to "local" stereoscopic cues are either simple or complex neurons. There are indications that disparity-tuned neurons are distributed in an orderly fashion similar to that found for orientation preference. In the squirrel monkey, neurons sensitive to retinal disparity were found in greater concentrations in the thick cytochrome oxidase-stained stripes. Occasionally, disparity-tuned cells were encountered in pale stripes (Hubel & Livingstone, 1987).

In conclusion, it appears that retinal disparity cues are extracted from the retinocortical input at an early stage, prior to the engagement of the processes that presumably lead to the extraction of complex stimulus properties, such as form, motion in depth, etc. Given that cortical binocularity is a prerequisite for stereopsis, disparity-tuned neurons play an important role in the operations that process stereoscopic cues and contribute to the perception of depth. Thus, tuned excitatory neurons appear capable of signaling differences in the relative distance of visual objects with a high level of precision. Far and near neurons are likely to be involved in coarse stereopsis and in providing signals of vergence error, which serve as triggers in the feedback mechanism that controls vergence eye movements (i.e., conjugate movement of the eyes in order to bring into focus a visual target that lies in front of or behind the point of fixation).

Visual association cortex: Area V3

Area V3 has received relatively little attention due, in part, to its anatomical location, which prevents easy access for electrophysiological and histological studies. V3 is generally considered a transitional or relay station in the extrastriate pathways. It receives a major input directly from layer 4B of V1 and also from area V2 (predominantly thick stripes). A significant proportion of V3 cells show direction selectivity (approximately 40%; Felleman & Van Essen, 1987), an observation consistent with the view that V3 serves as a major relay of magnocellular input. There are indications, however, that V3 may receive converging inputs that originate in non-magnocellular retinocortical pathways, via area V2 (Felleman et al., 1997). V3, in turn, projects to both areas MT and MSTd (which, as explained in more detail below, are critically involved in the analysis of visual motion) and area V4 (which is the main source of input to inferior temporal cortices). Area VP was originally considered as a subdivision of area V3, but recent histological data indicate that is has a distinct connectivity profile and is, therefore, considered as a separate area (Felleman et al., 1997).

The temporal visual pathway: Area V4

Area V4 occupies a zone of neocortex that extends from the anterior bank of the lunate sulcus to the posterior bank of the superior temporal sulcus (see Figure 21). Its posterior border is defined by areas V2 and V3, and its anterior border, by area MT and the posterior inferior temporal area (PIT). It receives dense projections from area V2 (arriving in layer IV) that originate primarily within thin cytochrome oxidase stripes and pale interstripes. It appears that adjacent cytochrome oxidase subcompartments within V2 thin stripes may project to non-overlapping V4 patches (Felleman et. al., 1997). As mentioned previously in this chapter, it is unclear if these histologically distinct subcompartments in

V2 comprise functionally distinct modules. Therefore, the functional significance of anatomically segregated input from V2 to V4 is not known. There is also evidence for converging input from both thin stripes and interstripes onto single V4 foci. This evidence is consistent with reports (based on optical imaging) of discrete V4 patches that become predominately active in response to either chromatic or oriented luminance contours, though preferential sensitivity to stimuli that contain chromatic oriented contours is very common (Ghose & Ts'o, 1994). Area V4 is considered to be the primary source of retinal input to the inferior temporal cortex, which is discussed in the next section. Area V4 probably contains a subregion, area V4a that receives most of its input from area V4 proper. The role of area V4 in many aspects of visual perception has been examined in detail and will be discussed below. Relevant visual capacities include, perception of fine details in visual stimuli (such as form, pattern, and texture perception) and color perception. In addition, the degree to which the responses of V4 neurons are modulated by attention has also been examined.

When tested with simple visual stimuli, such as edges and gratings, the majority of area V4 cells show similar response properties with complex neurons in areas V1 and V2, such as relatively narrow orientation preference (typically within 25° on either side of the best orientation). In fact, distinct cortical patches, similar to the iso-orientation columns in V1 and V2, have been found in area V4, as well (see Figure 54). The receptive fields of most V4 neurons lie within the central 30° of the visual field and are many times larger than those of V1 cells at similar eccentricities. Often, receptive fields extend into the ipsilateral visual field. Many V4 neurons show sensitivity to wavelength contrasts. The degree of selectivity of these cells to wavelength appears to be comparable to the spectral tuning displayed by cells in area V1 (Schein & Desimone, 1990). Unlike areas V1 and V2, where orientation sensitive and wavelength sensitive cells form separate clusters, many V4 cells display orientation as well as wavelength sensitivity (Desimone et al., 1985; Schein & Desimone, 1990).

In general, experimentally induced lesions confined to area V4 in monkeys produce only mild to moderate deficits in the discrimination of simple sensory cues such as orientation, and wavelength (Heywood & Cowey, 1987). Neither visual acuity nor the ability to discriminate objects on the basis of coarse features is significantly impaired. Other capacities, such as stereopsis, motion and flicker perception remain unaffected after V4 lesions (Schiller, 1996). These results can be easily accounted for on the basis of general principles of brain organization that were outlined in the Introduction. Electrophysiological evidence suggests that cells in several visual areas, at various hierarchical levels, show changes in firing rate when tested with simple stimuli such as spots of light of various wavelengths, edges, and gratings. It is then likely that in animals subjected to damage in area V4, spared visual cortex in areas V1, V2 and V3 is sufficient to support the ability to discriminate stimuli on the basis of their orientation or wavelength, and the ability to detect bar gratings across a wide range of spatial frequencies and contrasts. Another characteristic of many V4 neurons, the significance of which will emerge in the following paragraphs, is that they possess silent suppressive surrounds that may be as large as 30° in diameter. The response of these cells is suppressed by simultaneous stimulation of both the center and surround regions by light of the same wavelength, or by a grating of the same spatial frequency. Stimulation restricted to the surround region does not affect the cell's firing rate (Desimone et al., 1985).

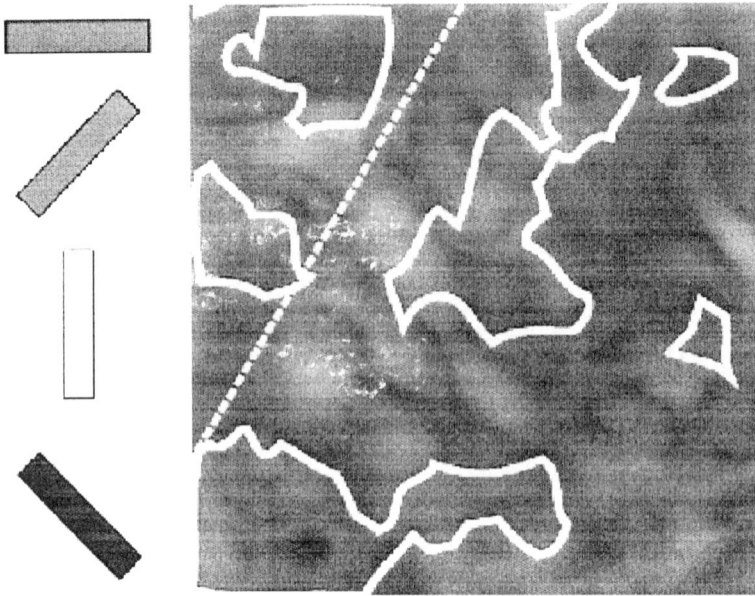

FIGURE 54.

Iso-orientation patches in area V4 revealed with optical imaging. Signal intensity in the map corresponds to stimulus orientation as shown with the bars on the left (the stimuli used to obtain these images were actually sinusoid gratings of different sizes). The dashed line marks the border with area V2. Areas enclosed by white lines probably contain cells that show strong receptive-field surround suppression. Note that, in contrast to area V1, the distribution of patches with orientation sensitive neurons bears no apparent relation with the distribution of cytochrome oxidase-stained patches as in area V1. (From Ghose & Ts'o, 1997, *Journal of Neurophysiology*, 77. Copyright 1997 by the American Physiological Society).

Recordings of single-cell activity in area V4 in primates suggest that this area is specialized for the analysis of color attributes and plays an important role in the perception of color (reviewed by Zeki, 1990). These reports indicate that a significant proportion of V4 cells are narrowly tuned to stimulus wavelength. However, estimates on the proportion of wavelength-selective cells vary widely across studies according to the criteria used to define selectivity. It appears that if the criterion is based exclusively on the shape of the cell's spectral tuning curve in response to monochromatic spots of light, the proportion of color selective cells in area V4 is not considerably higher compared to other visual areas. Assuming then, that simply signaling the wavelength of light reflected from surfaces is not what area V4 cells are uniquely specialized for, then what role does this area play in the perception of color? One property of V4 cells not found earlier in the visual pathway is sensitivity to wavelength in parts of the visual array located beyond the classical receptive field. The capacity to integrate wavelength-specific inputs from a relatively large portion of the visual field renders V4 neurons suitable for coding visual cues involved in context effects in color perception. One such effect is color constancy, described in more detail below. In order to capture the significance of this property, it is important to realize that wavelength sensitivity alone is not sufficient to sustain accurate color perception under natural viewing conditions.

It is rather difficult to realize intuitively the distinction between wavelength and hue (perceived color) because human observers do not possess an absolute sensitivity to wavelength: if they are asked to estimate the wavelength of the light reflected from a particular surface, they invariably associate a specific color to their sensation. Moreover, color perception shows an impressive immunity to variations in the actual wavelength composition of the reflected light. Thus, an observer continues perceiving a surface as red even when it is forced to reflect more short (green) and middle light wavelengths (blue) than long wavelengths (red).[11] It is largely due to this phenomenon (which is known as *color constancy*) that color plays such an important role in object recognition as a distinctive and relatively permanent object characteristic. The sensitivity of individual neurons to wavelength can be dissociated from their sensitivity to color, as perceived by the human observer, by varying the wavelength composition of the light incident upon multicolored surfaces (Land, 1974). Using these stimuli, Zeki (1983a) identified two distinct populations of wavelength-sensitive cells in areas V1 and V4. *Wavelength-selective* cells responded to the presence of a particular wavelength band in the light reflected from any part of a multicolored surface. For instance, cells selective to long wavelengths increased their firing rate when any portion of the stimulus that fell within their receptive field reflected a sufficient amount of long wavelength light, regardless of the perceived color of that surface. These cells also responded when their receptive field fell on a surface that reflected a large proportion of long wavelengths which, combined with smaller amounts of middle and short wavelengths, made the surface look white. In contrast, the responses of *color-coded* cells were tuned to the perceived color of a given surface. For instance, a color-coded neuron sensitive to red responded vigorously even when the area that fell within their receptive field contained a large amount of short and middle wavelengths and a smaller amount of long wavelengths. The same neuron did not respond to an area that reflected primarily long wavelengths when its perceived color was white. Therefore, these cells were sensitive to the color assigned by human observers to surfaces that were part of multicolored displays. Given that perceived color in such stimuli is also determined by the wavelength composition of light reflected from surrounding regions of the visual array, one can assume that color-coded neurons can integrate signals regarding the wavelength composition of the light reflected from areas outside the cell's classical receptive field. Thus, it is very likely that sensitivity to perceived color, as opposed to sensitivity to the wavelength of light reflected by a visible surface, may be a unique characteristic of cells in area V4, and that this property may form the basis for the perceptual phenomena of color constancy. Further, the sensitivity of neurons in area V4 to the wavelength and contour in parts of the visual field that lie outside the cell's classical receptive field, may play a role in perceptual phenomena of figure-ground segregation.

In humans, severe impairments in the perception of color, known as *acquired achromatopsia,*[12] are associated with damage to the caudal portion of the fusiform gyrus, which is located in the ventral surface of the occipital lobe (Damasio et al., 1980). In one particular case study, severe deficits in color discrimination were found following bilateral lesions confined to Brodmann's architectonic areas 18 and 19 (Rizzo et al., 1992). On the left side, the lesion was restricted to the visual cortical areas that lie below the calcarine sulcus. Inferior temporal and ventral occipito-temporal cortices (Brodmann's areas 37 and 20) were presumably left intact. Data from brain-damaged patients are consistent with findings from metabolic imaging studies with neurologically intact subjects. In a recent study (McKeefry & Zeki, 1997), normal adult subjects were presented with two alternating visual arrays, each a composite of the same visual elements as shown in Figure 55. In one of the arrays, the different visual elements varied in hue but were equal in reflected luminance (isoluminant chromatic stimulus). The second array contained luminance but not wavelength contrasts (achromatic stimulus). Subjects were asked to observe the stimuli while changes in regional brain oxygenation levels were measured using functional Magnetic Resonance Imaging. Regions that showed significant changes in tissue oxygenation included the occipital lobe (areas V1 and V2) and also the posterior part of the fusiform gyrus bilaterally. This data strengthen the proposed link between activity in ventral occipito-temporal cortex and the subject's engagement in tasks that involve the analysis of wavelength cues. Color constancy, on the other hand, may depend upon different areas than perception of color contrasts. Thus, specific deficits in color constancy were found after lesions were inflicted in anterior temporal regions, sparing the occipito-temporal cortex (Rüttinger et al., 1999).

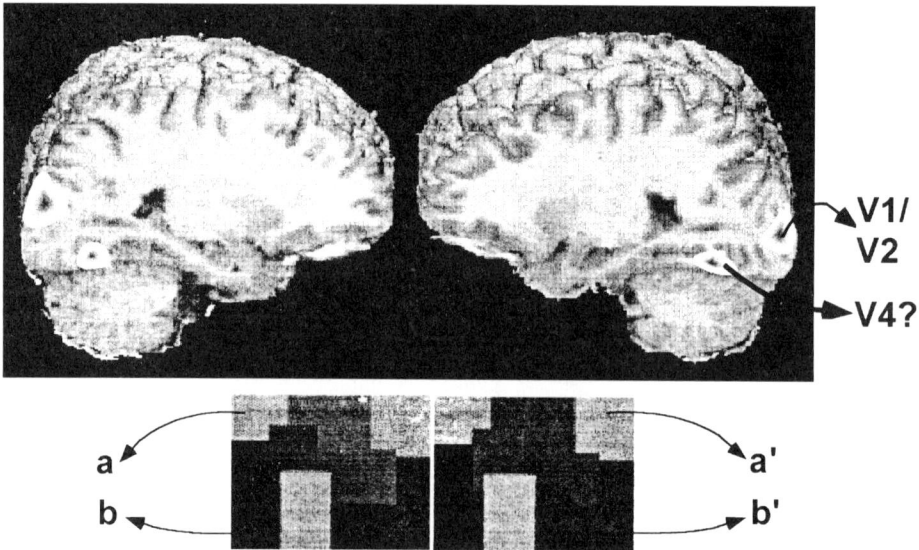

FIGURE 55.
Areas that showed significant changes in oxygenation levels (indicated by bright patches) in a representative subject during stimulation with alternating chromatic (lower left) and achromatic (lower right) Mondrian displays. The left hemisphere is on the right-hand portion of the composite. All elements in the chromatic display, like those labeled "a" and "b", had the same luminance, so that perception of the stimulus was supported only by wavelength-specific information. In contrast, only luminance cues were present in the achromatic display. (Based on data from McKeefry & Zeki, 1997).

Spatial attention appears to have a pronounced influence on the responses of V4 cells, a property that has not been found in cells earlier in the visual cortical pathway (Moran & Desimone, 1985). When monkeys were trained in a match-to-sample task, an effective stimulus (which had been determined earlier for each cell) elicited a good response when presented in an attended visual field location. The same stimulus, presented in an unattended location, and paired with a previously ineffective stimulus in the attended location produced little or no response. It should be noted that in the latter case, both stimuli were within the cell's classical receptive field. Moreover, it has been shown that the responses of cells in area V4 are often strengthened when the animal is engaged in a difficult orientation-discrimination task as opposed to a perceptually easier version of the same task (Spitzer et al., 1988). The effects of task difficulty may reflect the increased attention paid to the preferred stimulus presented within the cell's receptive field.

Finally, the contribution of neuronal activity in area V4 to the detection of complex visual contours remains an unresolved issue. Early reports have indicated that V4 lesions in monkeys cause severe deficits in pattern discrimination (Heywood & Cowey, 1987). Subsequent investigations employing more refined lesion and behavioral testing techniques suggest that deficits in pattern shape discrimination following V4 lesion are rather mild (Schiller, 1993). There is, however, evidence that V4 plays a major role in tasks involving the discrimination of target stimuli embedded in arrays of other more prominent stimuli (Schiller, 1993). The effect was seen regardless of the sensory cue (luminance, color at isoluminance, motion, or depth) that distinguished target from non-target stimuli. The perceptual capabilities probed by these tasks are highly significant ecologically, being crucial for the ability to detect, and identify, less salient environmental stimuli. Repeated training on the same tasks over a period of several days can lead to moderate behavioral recovery. It is notable, however, that V4-lesioned monkeys generally lack the capacity to generalize the regained skill to new spatial locations, suggesting a role of V4 in visual learning, a property that is even more prominent for inferior temporal areas (see below). V4 cells, with their unique receptive field surround properties (Desimone et al., 1985), may be part of the brain mechanism that supports perceptual capacities, like the ones reported by Schiller (1993), which rely on "figure-ground segregation".

The temporal visual pathway: Inferior temporal cortex (Area IT)

Functional properties of IT neurons
The inferior temporal regions in the macaque receive visual input from a variety of extrastriate areas, including the ventral occipito-temporal region (area VOT), the fundal superior temporal area (FST), and the superior temporal polysensory area (STP). They also receive input from the parahippocampal gyrus and area TG in the temporal pole (Felleman & Van Essen, 1991). Although a direct projection from area V2 thin stripes and interstripes has been found (Nakamura et al., 1994), area V4 is considered as the primary source of retinotopic input to area IT. The latter projects to the amygdala, the hippocampus via the entorhinal cortex, and to Brodmann's areas 46 and 8 in the frontal lobe. On the basis of several criteria (outlined in Chapter 1), area IT has been divided into several subregions, shown in Figure 56. Principal and alternate names of these areas are given in Table 1.

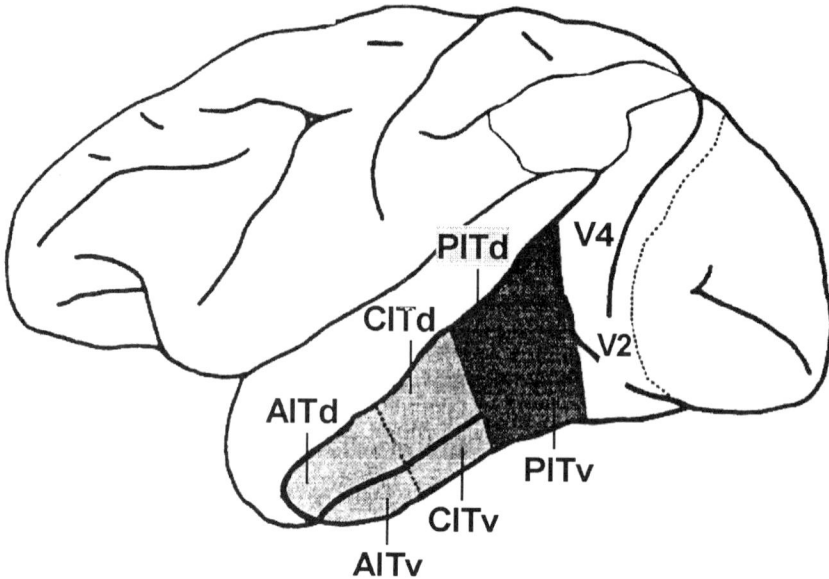

FIGURE 56.
Schematic drawing of the left hemisphere of the macaque brain showing the anatomical location of
inferior temporal areas. Alternate names for these areas are listed in Table 1. Areas buried within
the superior temporal sulcus are shown in Fig 21. Abbreviations; PIT: posterior inferior temporal,
CIT: central inferior temporal, AIT: anterior inferior temporal, PP: posterior parietal cortex. The
"v" and "d" suffixes indicate 'ventral" or "dorsal", respectively. (Based on data from Ungerleider
& Mishkin, 1982).

The posterior inferior temporal area (PIT) corresponds to the cytoarchi-tectonic subdivision known as TEO (Bonin & Bailey, 1947). V4 projections to this area (mainly
arising in layers 2 and 3 and terminating in layer 4) show remarkable convergence. Axons originating in several V4 patches, typically spaced 2-3 mm apart, often terminate
onto a single region in PIT (Felleman et. al., 1997). It is unclear at present if V4 patches
that give rise to these converging projections correspond to functional modules (as would
be the case if they were receiving input from different V2 compartments). The anterior
inferior temporal area (which corresponds to area TE of Bonin & Bailey, 1947) is subdivided into an anterior (AITa) and a posterior part (AITp), each further subdivided into a
dorsal and a ventral part. AIT probably corresponds to the cytoarchitectonic area TEp,
which receives a direct projection from area V4 and also from area TEO. AIT, on the
other hand, receives a direct projection from area TEO but not from area V4 (Baizer et
al., 1991). Inferior temporal areas project to several extratemporal regions including
some in the parietal and frontal lobes. Parietal areas (e.g. regions located inside the intraparietal sulcus, such as the lateral intraparietal area– LIP) receive extensive projections
from TEO. Consistent with these findings are recent reports that many LIP neurons show
considerable shape selectivity (Sereno & Maunsell, 1998).

Table 1. Nomenclature of inferior temporal visual areas.

Principal name *	Alternate name **
PITd	TEOd
PITv	TEOv
CITd	TEp or TEd
CITv	TEp or TEv
AITd	TEa or TEd
AITv	TEa or TEv

According to Felleman & Van Essen, 1991 and Seltzer & Pandya, 1989**.*

Projections from area TE target mainly prefrontal and medial temporal areas (Webster et al., 1994).

Electrophysiological studies using simple visual stimuli have shown that the majority of inferior temporal neurons typically have large receptive fields that include the fovea and often extend into the ipsilateral visual field (they may reach up to 65° in diameter). Convergence of inputs from multiple foci in area V4 may be responsible for the increase in receptive field size noted in this area (Felleman et. al., 1997). Visual responses are typically greater to stimuli presented near the fovea. IT neurons are not narrowly tuned to simple sensory cues such as stimulus orientation. There are conflicting views regarding the degree of selectivity displayed by IT neurons to more complex visual objects. One group of studies has identified cells in area IT and in the upper bank of the superior temporal sulcus (in areas TPO and PGa of Pandya & Yeterian, 1985), in the awake monkey, that respond with remarkable specificity to pictures of faces but show little activity in response to equally complex, biologically significant objects (Perrett et al., 1982). A second group of studies has shown that many IT cells show broad selectivity to a wide range of visual stimuli (for instance, Desimone et al., 1985).

The findings of the former group of investigations will be summarized first. A subset of the face-sensitive IT cells shows preference for the face of a particular individual (Perrett et al., 1984), whereas others appear to be sensitive to the general layout of faces and continue to respond despite changes in viewing conditions, such as retinal size, orientation, spatial frequency (i.e., blurring), wavelength, contrast, position, and viewing distance (Rolls & Baylis, 1986). Other neurons, however, are tuned to a narrow range of perspective views of a particular face: some cells are maximally sensitive to profile views, others to frontal views, and yet another group of cells responds to one profile view of the head (e.g., the left) and not the other (Perrett & Mistlin, 1989). According to one proposal neurons, which are sensitive to face identity regardless of perspective view, may be involved in signaling *object-centered* representations, whereas the activity of cells that prefer certain views and not others may encode *viewer-centered* representations (Perrett et al., 1989).

Many researchers suggest that these findings lend support to the so-called *neuron doctrine* or *sparse coding* theory (Barlow, 1972): the view that the coding of meaningful perceptual entities in the extrastriate visual cortex is based on the output of individual cells. These theories predict that individual cells should show great specificity for biologically relevant, complex visual stimuli. In its most strict version, this hypothesis postulates that the response of a relatively small number of neurons in a given area is sufficient for signaling the presence of a complex visual stimulus, such as the face of an individual. A fair amount of redundancy is thought to exist, in that entire groups of neurons have very similar stimulus preferences and, as a consequence, contribute equally to the discrimination of one object or face among other members of a given category or species, respectively (Perrett et al., 1989). A highly orderly hierarchical organization of projections within, as well as between, cortical areas is postulated in order to account for the increasing complexity of response preferences.

There are several problems with sparse coding theories. First, if the activity of neuronal aggregates that contain units with similar stimulus preferences is used to signal the presence of a particular stimulus, there simply wouldn't be enough neurons in cortical association areas to code the vast number of different perceptual entities (i.e., distinct objects, and living things at a great variety of sizes, contexts, visual contrasts, and perspective view) that can be recognized by the typical observer (Maunsell & Newsome, 1987). A second serious argument against the single-cell coding hypothesis is that cells in area IT and the superior temporal sulcus do not respond in an all-or-none fashion to the image of a complex visual stimulus. Although most neurons display preference for certain stimuli, they typically respond in a graded fashion to a range of stimulus attributes (e.g., Baylis et al., 1985). More recently, evidence acquired by means of a variety of electrophysiological techniques provides some support to the alternative to sparse coding, *population coding* theories. Next we turn to studies that examined the viability of population coding of complex stimulus attributes in area IT of the macaque.

Findings from studies using traditional analysis methods to examine the stimulus preferences of a single cell at a time are consistent with the view that individual neurons signal a particular *configuration* of visual attributes, and respond to a given face or object only when the latter contains the preferred combination of features. This position is strongly supported by observations that moderately complex, abstract patterns are sufficient to activate cells in AIT that were initially selected on the basis of their preference for the two-dimensional image of a real object (Fujita et al., 1992). These findings were obtained by first identifying a cell with a consistent preference for a real object (usually the picture of a plant or animal; see Figure 57). Next, the image was simplified using computerized image processing and the cell was tested again in order to determine the simplest effective image. The latter often consisted of a simple two-dimensional form or contained a combination of features (such as color, shape, and texture). By recording along oblique penetrations through the cortical mantle, Tanaka and colleagues discovered that cells with similar, yet not identical, preferences for such patterns tended to cluster together in vertical columns. These clusters had rather sharp boundaries and their average diameter matched the size of hypercolumns in area V1.

The notion that distinct subpopulations of inferior temporal cells code stimulus characteristics based upon the presence of one or more sensory cues is consistent with evidence that individual neurons in the anterior part of this area display broad tuning to a

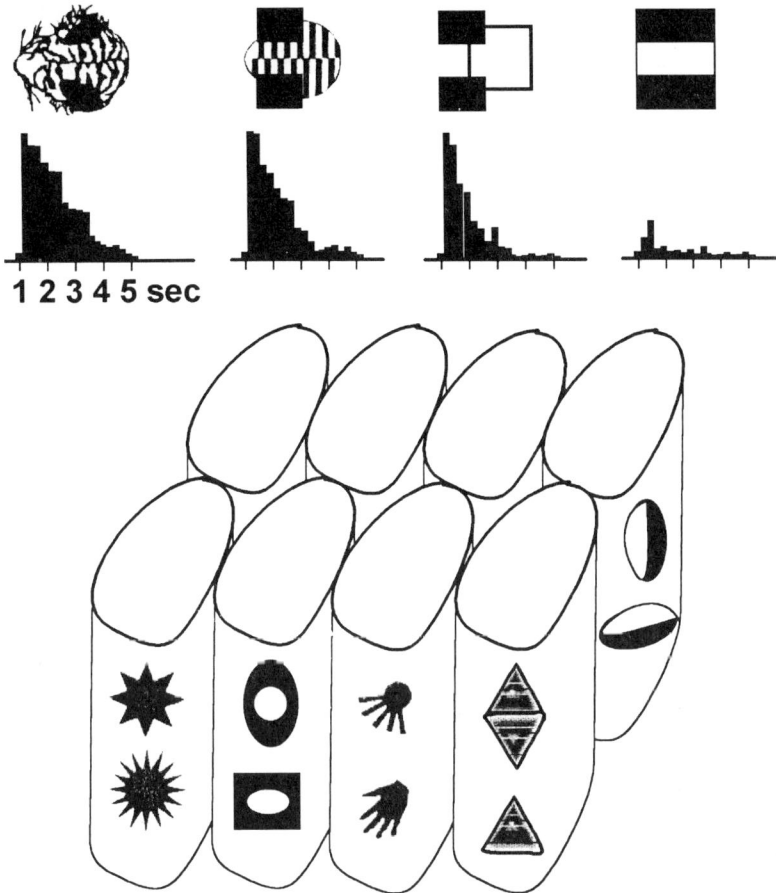

FIGURE 57.
One example of the stimuli used by Tanaka and associates to determine the critical features preferred by IT neurons. In this case, the drawing of the top view of a tiger head undergoes various stages of simplification (from left to right). Poststimulus response histograms, displaying the cell's average response over a number of repetitions of each stimulus are also displayed. (Based on data from Tanaka 1993.)

variety of geometrical shapes (Komatsu & Ideura, 1993). Other studies have revealed the sensitivity of IT cells to the general spatial characteristics of visual contours (Desimone et al., 1985).

Selectivity to complex patterns is more often found in inferior temporal regions. Moreover, many cells in this region show tolerance for a wide range of changes in the characteristics of the preferred stimuli (Miyashita & Chang, 1988). Therefore, it appears that a large population of inferior temporal neurons must be involved in signaling the visual image of even simple two-dimensional patterns. The discharge patterns of neighboring IT neurons display a moderate degree of intercorrelation. This was suggested by the multiunit analyses recently performed by Gawne and Richmond (1993). Taken together, these findings suggest that area IT, and especially its anterior part, is organized into hundreds of small modules that contain cells with partially overlapping preferences for visual patterns. When a real object is presented to the animal, a subset of these modules becomes simultaneously active. Synchronous activation of a particular set of modules is probably sufficient to signal the presence of a visual object.

The sensitivity of IT neurons (at the population level) to face stimuli has been examined in a series of studies by Yamane and associates these studies electrophysiological recordings from a large number of electrophysiological recordings from a large number of AIT cells were made while the animal was engaged in a discrimination task involving both familiar and unfamiliar faces. Virtually all the cells that were examined demonstrated broad tuning to each of the face stimuli. In other words, the activity of individual neurons was clearly not sufficient to signal the presence of any single face. Moreover, these findings were consistent with the view that each cell participates in the representation of many different faces, rather than the face of a single individual. In summary, the response properties of neurons in the anterior inferior temporal area indicate that different visual attributes, such as color, shape, and pattern, become integrated and simultaneously coded in the activity of these cells (Komatsu & Ideura, 1993). The sensitivity of these neurons to more than one visual property, combined with their large receptive fields, would make them suitable for signaling the presence of objects on the basis of various combinations of visual features.

The reports of face-selective cells in area IT and adjacent superior temporal sulcus regions raised the possibility that these areas are homologous to the regions involved in face agnosia in humans. Face agnosia, or *prosopagnosia*, is an acquired inability to recognize faces (for a more detailed description see Box 4). In humans, this disorder is associated with damage to the ventromedial occipito-temporal neocortex (Brodmann's area 37) and the adjacent parahippocampal gyrus (Damasio et al., 1990). To examine the correspondence between these regions and inferior temporal cortex in primates, researchers compared the performance of macaque monkeys, who had been subjected to bilateral removal of the neocortex surrounding the superior temporal sulcus, with the performance of four well-studied prosopagnosic patients (Heywood & Cowey, 1992). Contrary to expectations, the lesioned animals did not perform significantly worse than the normal controls on a number of tasks in which prosopagnosic patients displayed severe deficits. Although the lesioned group showed a mild impairment in a two-choice match-to-sample task, their discrimination defects were not only specific to faces but extended to other objects as well. Recognition of face identity, regardless of orientation, was unimpaired. Moreover, their ability to discriminate faces on the basis of their familiarity and their significance for the animal (a test analogous to a face identification task for humans) was not significantly worse compared to the unoperated control animals. The prosopagnosic patients, on the other hand, were severely impaired in tasks that required face identification and face discrimination on the basis of familiarity. These findings have shed some doubt on the claim that prosopagnosia is the result of damage to a region homologous to areas surrounding the superior temporal sulcus in primates. It is possible that the regions, which are indispensable for the ability to perceive facial identity, reside in more ventral temporal lobe areas in primates.

Bilateral ablations of the inferior temporal cortex can also lead to deficits in pattern discrimination (Bolster & Crowne, 1979).

BOX 4. SUBTYPES OF FACE AGNOSIA

Pure Associative
- Severe deficit in recognizing familiar and famous face identities and in learning new identities.

- The impairment is restricted to the visual modality; patients can recognize a person by gait, voice, etc.

- The impairment is not due to visual field defect, or to poor acuity or contrast sensitivity (Rizzo et al, 1986; 1992).

- Recognition of other objects at the basic level may also be impaired (e.g., the patient identifies a cat as an animal but he confuses a cat with a lion).

- Often accompanied by color deficits (central dyschromatopsia).

Amnesic Associative
- Inability to recognize face identity in every modality; probably associated with a severe deficit in episodic recall.

- Covert recognition is often observed (in priming, electrical skin measures, and oculomotor scanning patterns).

Partial Apperceptive
- Associated with a deficit in integrating all parts of a visual stimulus into a unique percept.

- Restricted to the visual modality.

- Deficits in complex visual processing are present. Patients may show poor ability for matching visual stimuli across different views, drawing items that they cannot identify, and matching them to sample.

The severity of these deficits, however, has been a matter of dispute. It appears that although pattern discrimination deficits can be severe immediately following the lesions, the animals may gradually learn to distinguish visual patterns and objects after prolonged training. Evidence from a number of studies suggests that monkeys with IT lesions suffer from a true deficit in the perception of two-dimensional patterns. Often, however, they learn to use other visual cues, such as stimulus size or luminance, in order to perform visual discrimination tasks. The extent of the ability to utilize less salient cues, which are not used regularly by normal monkeys, probably depends upon the contribution of other visual areas that remain intact.

Perhaps the most impressive evidence to date that the activity of IT reflects conscious perception, and not merely preference for a particular configuration of visual elements, comes from the study of binocular rivalry. The latter refers to the (reported) perceptual dominance of one of two visual stimuli that are presented dichoptically (i.e., only one stimulus to each eye). The construction and presentation of the two stimuli precludes binocular fusion.

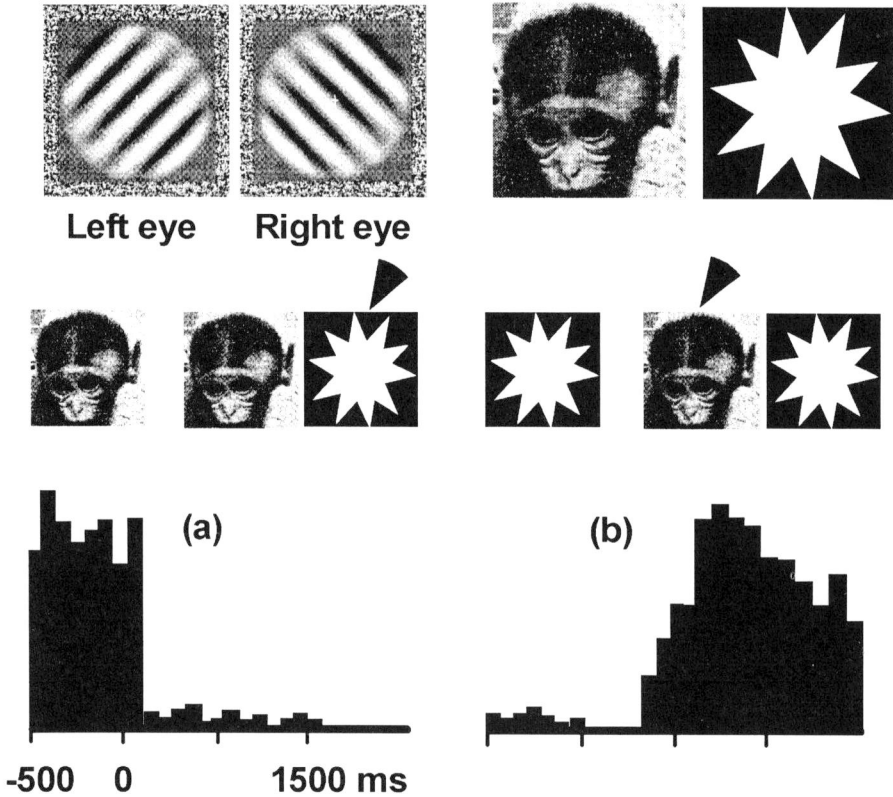

FIGURE 58.

Examples of visual stimuli used in perceptual rivalry experiments. *Top left*: two grating patterns of orthogonal orientations used in electrophysiological studies in areas V1, V2, V4, and MT. *Top right:* Stimuli used in studies of IT neurons. *Bottom:* Typical responses from an IT neuron. First, a variety of stimuli are presented monocularly until a preferred stimulus is found, as indicated by the cell's vigorous discharge (initial portion of each poststimulus response histogram). Then the preferred and a non-preferred stimulus are presented dichoptically (one to each eye) and the monkey is trained to report, by pulling a lever, perception of the geometrical pattern (during condition "a") or perception of the face (condition "b"). The perceptually dominant stimulus is indicated by an arrowhead. (Examples of stimuli are reprinted from Logothetis, 1998, Single units and conscious vision. *Philosophical Transactions of the Royal Society London B., 353*, pp. 1801-18. Copyright 1998 by the Royal Society. The histograms are based on data from Logothetis, 1998.)

Instead, observers report a fluctuating percept with alternating "perceptual dominance" of each pattern. The simplest stimuli used to reproduce this phenomenon are grating patterns of orthogonal orientations, like those displayed in Figure 58. This paradigm permits the study of neurophysiological correlates of changes in conscious perception in the absence of changes in stimulus properties. It is possible to train monkeys to indicate which of the two patterns is perceived at a given point in time while microelectrode recordings of neuronal activity are performed.

In a series of experiments (reviewed by Logothetis, 1998) it was found that the responses of only a small proportion of cells in areas V1, V2, V4, and MT correlated with the perception of complex stimuli, including faces and geometrical patterns (see Figure 58), as indicated by the animals' behavioral response. In addition, a sizable proportion of neurons in areas V4 and MT responded more to perceptually suppressed stimuli. In these experiments, it was found that nearly all neurons studied in area IT changed their activity depending upon the perceptual dominance of a preferred visual stimulus. That is, neurons responded vigorously to a normally preferred stimulus when it "dominated" perception. When the same stimulus was not the dominant one, the cells' response decreased significantly. Taken together, these findings suggest that neuronal processes that take place in the inferior temporal cortex more closely reflect the output of psychological operations that give rise to the conscious awareness of visual input than any other area in the visual system.

In conclusion, area IT and adjacent regions in the ventral bank and the fundus of the superior temporal sulcus appear to be involved in visual operations that lead to the discrimination and recognition of complex visual patterns, including such biologically significant stimuli as faces. There is a continuing debate regarding the nature of the neural code that is used by the central nervous system for performing such operations. At present, existing evidence favors a neural code based on the pattern of activity distributed within large subpopulations of visually responsive neurons. However, a direct empirical demonstration of population coding in the visual system has been difficult to achieve, in contrast with other cerebral systems such as the motor system (e.g., Georgopoulos et al., 1986). Moreover, researchers have yet to demonstrate that a population-based neural co-code is sufficient to support complex perceptual decisions.

Evidence for a role of area IT in associative learning

In addition to causing deficits in the discrimination of familiar visual objects, temporal lobe lesions impair the ability to learn to discriminate new objects and patterns and to distinguish between familiar and novel objects (visual memory). Typically, learning in other modalities remains unaffected. Electrophysiological findings are consistent with the lesion data. For instance, single unit recordings suggest that the activity of neurons in area IT may reflect the formation of associative relations between complex visual patterns. This possibility was explicitly tested by Sakai and Miyashita (1991) in two macaque monkeys. First, neurons were identified on the basis of their strong preference for at least one out of a large set of visual patterns. Next, the animal was trained to form associations between pairs of abstract geometrical patterns that bear no systematic visual similarity to each other (see Figure 59). As expected, many cells responded vigorously to

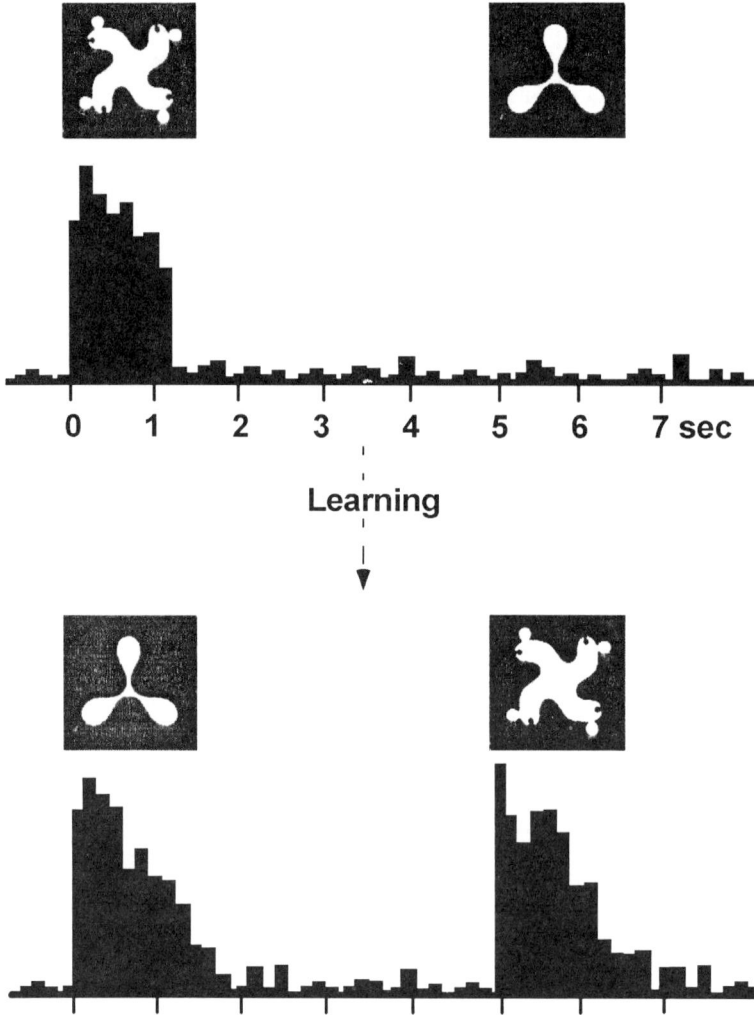

FIGURE 59.

The paradigm used to demonstrate the associative properties of IT neurons. The animal learns to associate two computer-generated pictures one of which (top left) elicits a strong response from the neuron shown here. After the association is formed the initially ineffective stimulus elicits a strong response from the same cell. (Reprinted by permission from *Nature*, *354*, pp. 152-155. Copyright 1991, Macmillan Magazines Ltd.).

more than one pattern. Interestingly, preferred stimuli were significantly more likely to belong to "learned" pairs, suggesting that the cell's preference had been "shaped" through the repeated exposure to the paired stimuli rather than being determined by visual similarity. These *"pair coding"* neurons were clustered together within a small patch in AIT.

The presumed "associative" property of IT neurons may underlie the sensitivity of some cells to object-centered (Marr, 1982) or prototypical representations (Perrett et al., 1989). According to this theory (Miyashita, 1993), different views of the same object are typically perceived in close temporal contiguity during everyday visual experience, thus allowing the formation of long-term associations between the different viewer-centered representations. Indeed, IT neurons have been identified that respond similarly to different visual patterns that were paired together during repeated joint presentation to the animal (Miyashita, 1988).

A second group of neurons encountered in area AIT increased their firing rate during the delay between the first and the second stimulus in each pair (termed *"pair-recall"* neurons). Each of these cells displayed a sustained increase in firing rate only in response to paired-associates of preferred patterns, in other words, to stimuli that failed to excite the neuron prior to training (see Figure 60). According to one interpretation, pair-recall neurons may be part of the neural mechanism involved in the formation (or the "retrieval") of the neural "representation" of visual images. The "associative" properties of neurons in area AIT are consistent with the fact that this area is the final station in the pathway that interconnects visual cortices with medial temporal lobe (MTL) structures, which are known to be involved in memory functions (e.g., Squire, 1986).

Like neurons in area V4, the responses of many IT cells are modulated by spatial attention. To examine these properties, monkeys were trained to maintain fixation on a central spot of light while a variety of stimuli were presented peripherally. When the animal was trained to attend to the shape of the peripheral stimuli in order to make pattern discriminations, an increase in the firing rate of individual IT neurons in response to the onset of these stimuli was observed (Richmond & Sato, 1982, cited in Richmond et al., 1983). In contrast, when the animal was trained to perform a simple luminance discrimination task (detecting a reduction in the brightness of the peripheral stimulus) the responses of IT neurons either remained unchanged or even became suppressed (Richmond et al., 1983). Attention-enhancement effects were also noted by Moran and Desimone (1985) in a match-to-sample task involving color and shape discriminations.

It may be recalled that attention effects involving cells in area V4 occurred only when both the effective and the ineffective stimulus were presented within the cell's (relatively small) receptive field. Suppression of the response to an unattended effective stimulus could be prevented if the focus of attention was shifted to an irrelevant stimulus outside of the receptive field. Neurons in IT generally show a similar response pattern. However, given the fact that the receptive fields of IT cells include a large portion of the contralateral, as well as the ipsilateral visual field, these neurons increase their firing rate in response to a preferred stimulus almost anywhere in the visual field as long as that stimulus is attended.

Single-cell data have not yet provided comprehensive answers regarding how different aspects of visual/semantic knowledge is represented in the cerebral cortex. In the next section, we present some of the most prominent accounts of how semantic knowledge is represented in the human brain. These models were constructed on the basis of a very different approach: the detailed examination of the residual capabilities and deficits in brain damaged patients. Two of the properties of neurons in visual association cortices are particularly relevant to some of these models: 1) population coding and the related

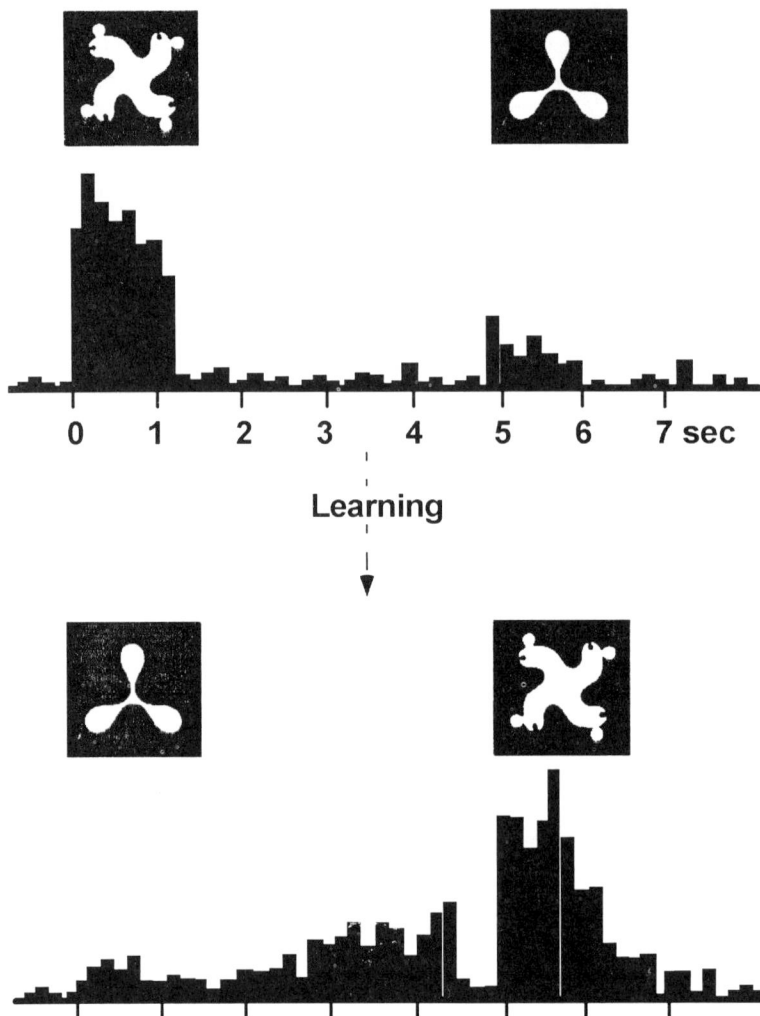

FIGURE 60.
Typical response profile of a "pair-recall" neuron in IT. Following extensive training in associating the two pictures, the initially ineffective stimulus elicits activity during the interstimulus interval. Notice that discharge rate slowly increases after the presentation of the first stimulus (lower display) up to the onset of the second stimulus, which is associated with another increase in firing rate. (Reprinted by permission from *Nature*, *354*, pp. 152-155. Copyright 1991, Macmillan Magazines Ltd.).

notion of distributed representations and, 2) the capacity to form associative relation between sensory cues. These properties are consistent with the functional requirements of one of the proposed models for the system that subserves visual recognition (Damasio et al., 1990; Komatsu & Ideura, 1993).

Neuropsychological models of "semantic representation"

Visual agnosia is traditionally defined as an impairment in the recognition of visual objects that is not attributable to sensory defects, mental deterioration, attentional deficits, or a disturbance in object naming (Bauer, 1993). For more than a century, descriptions of recognition deficits associated with brain injury have formed the basis for most neuropsychological theories on the organization of semantic knowledge. One of the most influential models developed in order to account for the disorders of agnosia can be traced back to Lissauer (1890; cited in Bauer, 1993, p. 216). This model proposes that object recognition has two main components: *apperception* and *association*. Apperception is the process of combining the visual features of an object into a single percept. Association involves linking a visual percept to previous experiences, and was thought to form the final step prior to consciously perceiving an object as a meaningful entity. Accordingly, *apperceptive visual agnosia* is characterized by deficits in the perception of the salient features of visual forms leading to pervasive difficulties in the recognition of familiar objects presented in the visual modality. The ability to discriminate objects based on shape (as in comparing two rectangles matched for total surface area but with different relative side lengths) is also impaired, and patients often rely on other features such as color, if available. Apperceptive visual agnosia is usually defined by the following signs (Benson, 1989):

- Difficulty in discriminating, naming, copying, or recognizing visually presented objects, including difficulty to identify familiar faces;
- Inability to draw or construct models of objects from memory or copy them;
- Naming, copying, or recognizing ability is intact for objects presented in other modalities. For instance, patients can name objects that they are allowed to manipulate, and line drawings of objects that they can trace with their hand. Also they can recognize a person by their voice;
- The ability to identify color and orientation of visual stimuli and to perform brightness discriminations is relatively intact.

Pure cases of apperceptive agnosia are rare, with the majority of agnosic patients displaying perceptual and cognitive deficits as well. Moreover, precise lesion localization in the reported cases is lacking, although, given the nature of the behavioral deficits, one would predict involvement of extrastriate visual areas primarily in the temporal pathway. A disturbance in associative processes, on the other hand is considered the hallmark of *associative visual agnosia*. A key feature that distinguishes the latter from apperceptive agnosia is the preservation of the ability to discriminate and copy visually presented objects and forms. Naming to visual confrontation, recognition, and appreciation of meaning in the visual modality are severely impaired.[13] Although the majority of reported cases are associated with bilateral lesions, unilateral left or right hemisphere damage alone can cause signs of associative agnosia (McCarthy & Warrington, 1986). Typically, the extent of the lesion is such that impairments in reading (alexia) and face recognition (prosopagnosia) are also produced.

A central assumption in this and other more recent models of visual recognition (e.g., Ellis & Young, 1988) is that perception and memory are mediated by separate mechanisms. For instance, Ellis and Young's (1988) "box" model implies that visual recognition is the final outcome of a hierarchically organized sequence of operations that become engaged after the extraction of simple visual features. Thus, visual input first produces a viewer-centered description of the object, which essentially is a two-dimensional representation of the object's spatial layout from the observer's point of view. Additional computations are performed before a three-dimensional, object-centered representation[14] can be constructed. Subsequently, object *descriptions* are compared with object *representations,* termed "object recognition units". The latter contain a structural (i.e., visual) description as well as semantic information regarding the object's functional attributes (such as its usage, its relation to other object, etc.). A successful match can trigger recognition, which in turn, is a prerequisite for the retrieval of the object's name in the form of a phonological representation. A controversial issue among the proponents of this general model is whether the structural descriptions of visual objects are stored separately from their respective semantic representations (Riddoch et al., 1988). These and other similar proposals have strongly influenced the clinical subcategorization of visual agnosia syndromes. An example of a recently proposed classification scheme for a specific form of impaired object recognition, face agnosia, is outlined in Box 4 (Damasio, 1990; see also Bauer, 1993).

The models outlined thus far have one fundamental property in common: they assume that the system that subserves object recognition is organized in a *modular* fashion. In other words, they postulate that object recognition consists of several component processes that operate relatively independently from each other and, perhaps more importantly, are anatomically dissociable. Proponents of such models traditionally seek supporting evidence in data from lesion (case) studies and in the detailed analysis of selective deficits and residual abilities of brain damaged individuals (e.g., Shallice, 1988). For traditional modular or *discrete-stage* models, signal transmission between hierarchically organized processes or operations proceeds in a strictly serial manner.

More recent accounts of visual recognition have introduced the notion of cascade activation, by analogy with interactive-activation models of information processing in artificial (computer) networks (McClelland & Rumelhart, 1981). In traditional discrete-stage models, access to information stored within each stage occurs independently of access to information in previous or later stages. Traditional models predict that experimental manipulations of certain stimulus characteristics should selectively affect the outcome of the operation performed in one stage without interfering with other operations. In *cascade* models, on the other hand, a given stimulus typically activates multiple representations, even at the early stages (for instance at the level at which the visual structural description of an object is complete). Each of the activated representations can in turn activate multiple representations at subsequent levels in parallel. In contrast to discrete-stage models, it is not necessary that the execution of an operation, which "belongs" to an early stage, be completed before the initiation of operations that belong to higher processing levels.

Other applications of the notion of modularity in neuropsychology
Another aspect of the concept of modularity is the distinction between *semantic* and word-like (or *lexical*) representation -- the latter being a prerequisite for object naming. The prototype for contemporary models in this area can be found in the work of Wernicke (1874). He proposed that the act of naming a visual object involves the "transmission" of a visual sensory image from visual to auditory association areas where a word-sound image is evoked. The latter is then "conveyed" to motor areas where it activates an articulatory program. Elaborating this hypothesis, Geschwind (1965) hypothesized the

existence of a higher-order association area that controls the functional connections between lower-order association areas. He placed this region in the inferior parietal lobule, which receives extensive input from most of the surrounding association cortex, including visual, auditory, and somatosensory areas. According to this model, the semantic attributes of an object, which are closely linked to its physical characteristics, are extracted in the visual association areas. The object's name is "stored" in posterior auditory association regions (such as Brodmann's area 22). The role of structures in the inferior parietal lobule is to mediate functional associations between separate sources of modality-specific stored representations.

Damasio and his associates reported data from lesion studies consistent with the notion that separate systems mediate lexical and visual/semantic knowledge (Damasio et al., 1990). Twenty three patients with focal brain lesions (eight in the right hemisphere, eleven in the left hemisphere, and four with bilateral damage), and two groups of neurologically normal control subjects (young and older adults) were presented with 320 pictures of various objects. In one condition, they were asked to name each stimulus. In the second condition, they were asked to provide a verbal description of the depicted object. In addition, their knowledge regarding the physical and functional characteristics of each object was probed with a series of questions. Closer examination of the patients' MRI data revealed that damage to the inferior temporal cortices (i.e., cytoarchitectonic regions 20, 21, and 37) was sufficient to cause severe recognition deficits for members of the *category* "living things" (see Figure 61). The subjects' ability to recognize inanimate objects (i.e., tools and utensils) was not different from the average control subject. On the other hand, damage to the left anterior temporal region impaired object naming but not recognition per se.

Yet another application of the concept of modularity in neuropsychology is in reference to the organization of object representations *within* the semantic system(s). A number of single-case studies have revealed patterns of category specific impairments of semantic knowledge. For instance, Warrington and Shallice (1984) described two patients recovering from herpes simplex encephalitis who showed very poor identification performance for members of the superordinate categories of "animals", "fruits", "vegetables", and "food". The same patients could identify inanimate objects with a fair degree of accuracy. A similar pattern of identification performance has more recently been reported by Sartori and Job (1988) in a patient with bilateral anterior temporal damage. Another patient who had suffered bilateral brain damage due to herpes encephalitis was found to have difficulty in the recognition and identification of members of various categories of living things compared to inanimate objects (Young et al., 1989). In that study, the influence of category typicality and item familiarity on category-classification performance was examined separately. The results showed that these factors affected the patient's ability to correctly classify living things to a greater extent than his ability to match verbal labels of categories to names of inanimate objects.

In order to account for the category-specific deficits in object recognition and naming displayed by a number of brain damaged patients, discrete-stage models have postulated the existence of separate "semantic systems based on functional specification [which] might have evolved for the identification of inanimate objects" (Warrington & Shallice, 1984). Cascade models, on the other hand, have emphasized the potential role of structural similarity between objects that belong to the same category. A fundamental assump-

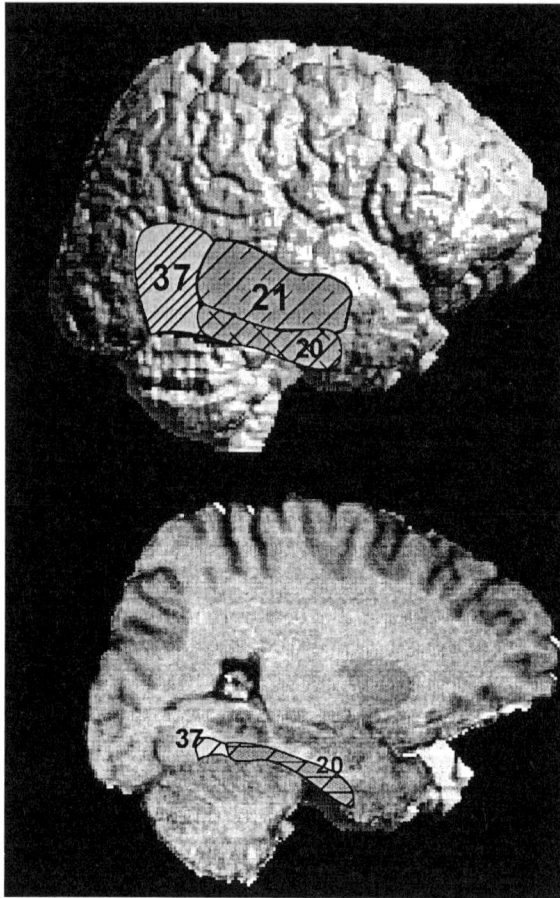

FIGURE 61.
The inferior temporal region in the human bran comprises cytoarchitectonic areas 20, 21, and 37. Unilateral or bilateral damage to this region can cause object recognition deficits (Based on data from Damasio et al., 1990).

tion in these models is that a given visual stimulus typically activates many semantic representations simultaneously. The latter correspond to objects that share a number of visual features. The higher the structural similarity between members of a particular superordinate category, the larger the spread of activation within the "semantic system". Lesion-induced impairments in the ability to identify a particular object could arise in two ways. First, the lesion may directly affect the "stored" representations of visual features that form the core structural description of a given category (Sartori & Job, 1988). Alternatively, the lesion may simply increase the amount of intrinsic noise in the process that mediates access to the semantic representation of a given object from one or more activated representations in the structural description system (Humphreys et al., 1990). Higher noise levels in this process could increase the likelihood for the activation of an inappropriate representation at a later stage. Naturally, recognition deficits are expected to be more severe and more common for members of categories that share a large number of common visual features, than for members of categories that display relatively fewer common features. This hypothesis predicts less severe deficits in the ability to identify objects as members of a superordinate category (e.g., animals, tools, etc.) because semantic information alone (as opposed to visual characteristics) would be sufficient to clarify the object's identity.

Consistent with the claim that the severity of visual recognition deficits is inversely related to the degree of structural similarity between members of a particular superordinate category, is the finding that some patients suffering from visual agnosia are capable of identifying certain "atypical" members of a given category, despite severe category-specific recognition impairment. For instance, one patient could identify an elephant and a giraffe, although she consistently failed to recognize more "typical" members of the category of animals such as a lion and a dog (Damasio et al., 1982). In summary, recent accounts of "category-specific" deficits have started to move away from traditional notions of modularity in the storage of semantic knowledge in the brain. A key concept in recent approaches to this problem is the hypothesis that "representations" of visual objects consist of numerous elements (each corresponding to a different feature) which: 1) have been linked together through the process of association[15] and, 2) are shared by many members of a given object category (and occasionally by less typical members of other categories).

An integrated model of the brain mechanism responsible for learning and remembering "semantic categories"

Recently, a model of visual object recognition and naming has been reformulated by Damasio and his associates (Damasio, 1989, 1990; Damasio et al., 1990), acknowledging the role of visual similarity between members of a category that was originally proposed by Humphreys (Humphreys et al., 1990). Damasio, however, suggests that structural characteristics are but one of several factors that determine the extent of the impairment in visual recognition abilities. Other factors include the sensory modality (or modalities) through which the individual has perceived a given object, the frequency of past encounters with a particular entity, and its value for the observer. Entities that have been coded in more than one modality are expected to be more resistant to recognition impairments. For instance, the ability to recognize a kitchen utensil that one uses on an everyday basis, and has perceived through multiple modalities (e.g., visual and somatic sensation) is more likely to be preserved in cases of visual agnosia than the ability to recognize a specialized tool that the observer has seen only in a picture.

According to Damasio (1989), the object recognition system is composed of a number of anatomically distinct subsystems. Patterns of activity distributed in neuronal assemblies located within primary and low-order association cortices (i.e., modality specific neocortex) reflect certain visual features of an object (called "feature fragments"). Interconnections between modality-specific areas (or *local convergence zones* such as areas V2 and V4) serve to integrate the representations of component features into meaningful entities. Visual areas V2 and V4 are classified in this category. The strength of these cortico-cortical connections (which are often called *connection weights* by analogy to artificial network systems) is reinforced by the repetition of relevant perceptual events. Neuronal assemblies in modality nonspecific brain regions, which have been termed *high-order convergence zones*, regulate the connections between local convergence zones. Their primary role is to ensure that a stimulus perceived through a single modality (e.g., the picture of an object) will activate related representations in other modality-specific areas that were established during previous encounters with that object. Feedback projections from high-order to local convergence zones play an important role in this process. An important feature of this model is that it does not make a distinction between perception and memory. Thus, the same neural elements that become activated when a particular object is first perceived participate in its retrieval at a later time in the absence of any related sensory input.

The high-order convergence zones are presumably located in tertiary association areas, and also in limbic cortices (which include medial temporal structures such as the hippocampus). Neocortical as well as paleo- and archeocortical limbic structures and certain parts of the basal ganglia may also contain "amodal" convergence zones. The model postulates that the response of neural units in these convergence zones is neither modality- nor stimulus-specific in the traditional sense. Rather, these units regulate the strength of the interconnections between modality-specific neural elements which, when simultaneously active, give rise to a particular perceptual experience. Therefore, they should not be confused with the so-called "grandmother" or "cardinal" cells (Barlow, 1972) that play an important part in sparse coding models. The role attributed to high-order convergence zones in Damasio's model leads to the intriguing prediction that damage to these cortices should cause recognition deficits predominantly for concepts that cannot be easily coded on the basis of physical characteristics alone. This also applies to the members of a superordinate taxonomic category that show a high degree of structural similarity to each other. For instance, most classes of living things (such as *land animals*, and *birds*) are characterized by a higher degree of visual ambiguity and featural overlap among their members, than the superordinate category of *tools*.

At the microscopic level, evidence that IT neurons show some intriguing associative properties (Miyashita, 1993) is consistent with Damasio's proposal. However, it should be noted that, thus far, IT neurons have only been tested for their capacity to reflect associations based on a limited set of visual features. At present, data concerning intermodal associative properties are lacking. A more rigorous test of Damasio's model would entail recording simultaneously from neurons in different cortical areas that respond predominantly to stimuli in different modalities. Target cells should show preference for different attributes of a given stimulus, or alternatively to conceptually unrelated stimuli that have been associated through conditioned learning. The model predicts that electrical stimulation of convergence zones should cause synchronous activation at these cortical sites. Recently, we have obtained evidence from in vivo macroscopic imaging of neuronal activity in humans that is consistent with this model. The functional brain imaging method used in these studies was Magnetic Source Imaging, a non-invasive technique that allows mapping of the distribution of activity changes in large neuronal populations in real time after the presentation of an external stimulus (such as a visual pattern in the context of a recognition task). We have observed activation of primary and secondary visual association cortices both early (within 150 ms after stimulus onset) and later (between 300 and 700 ms post-stimulus onset), that is, both before and after activation of higher order association areas in the same individual (Papanicolaou et al., submitted).

The parietal pathway

Cortical regions in the parietal lobes have long been thought of as essential parts of a system involved in visuospatial analysis (Ungerleider & Mishkin, 1982). Electrophysiological research in primates (e.g., Andersen, 1989) and the study of brain-damaged patients (Zihl et al., 1991) has revealed that parietal regions are also involved in the perception of visual motion. These perceptual capabilities play a key role in the guidance of eye and hand movements toward visual targets (Andersen, 1989; Dursteler & Wurtz, 1988; Gnadt & Andersen, 1988). Neurons in these regions show sensitivity and often selectivity to very diverse stimulus attributes, including visual motion, distance and 3-dimensional shape. Some units respond selectively to targets in certain spatial locations, whereas other cells are tuned to specific eye-movement trajectories. Many parietal neurons are multimodal and respond to visual, somatosensory and vestibular stimulation. In addition, there are dense reciprocal connections between parietal and premotor cortices, which are involved in the organization of reaching and grasping movements. These findings suggest that the parietal cortex is involved in "constructing" the layout of surrounding space, and in identifying the precise location of potential targets for eye, head, and hand movements.

In addition to signaling aspects of the visual array that can be used for motor planning, areas along the parietal pathway appear to be involved in purely perceptual functions, such as the appreciation of spatial layout and the perception of visual motion. Two areas in particular, the middle temporal (MT) and the medial superior temporal (MST), have been investigated extensively for their role in complex visual motion perception.

Physiological properties of cells in the middle temporal area (MT)

The middle temporal area is located in the dorsal bank of the superior temporal sulcus (see Figure 21). Neurons in this area have relatively small receptive fields (i.e., between 3-$10°$ in diameter within the central $10°$ of the fovea) that increase in size with eccentricity (Tanaka et al., 1986). The most prominent characteristic of cells in this area is their sensitivity to visual motion. The majority of neurons are tuned to a narrow range of directions of moving patterns. Most neurons show, in addition, suppression of firing when movement in their receptive field occurs in the anti-preferred direction. The majority of cells are tuned to a narrow range of stimulus velocities for both smooth and apparent motion.[16] Typically, their firing rate is reduced when the speed of apparent motion falls outside of a narrow range of preferred velocities. A process of active response suppression appears to be responsible for the fact that the discharge rate of many MT cells is dramatically reduced when the velocity of apparent motion falls below a particular value. This phenomenon is observed even for motion in the cell's preferred direction.

Like neurons in area V4, the response of many MT cells is affected by visual events that occur outside of the classical receptive field. Thus, a pattern that moves in the anti-preferred direction outside of the borders of the receptive field often suppresses the cell's response to motion in the preferred direction that occurs within the cell's receptive field. Other cells show maximum suppression when the direction of motion in the surround matches that of the central array, as shown in Figure 62 (Allman et al., 1985). These neurons are capable of signaling the exact location of motion discontinuities within the visual field. With their capacity to respond differentially to motion within the receptive field and to motion in the background, these cells may play a role in perceptual phenomena involving figure-ground segregation.

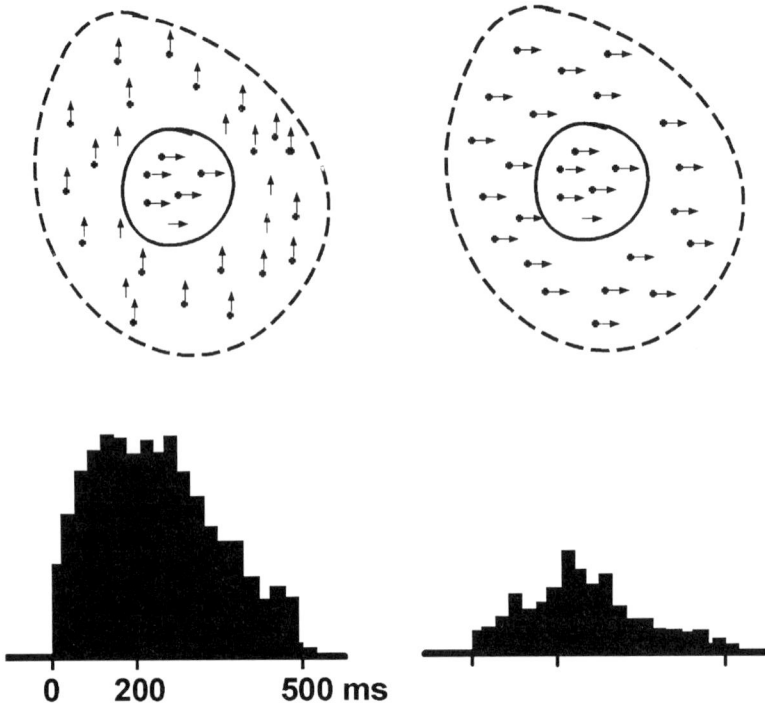

FIGURE 62.
Reduction in firing observed in a MT neuron when the moving elements in the "silent" receptive surround move in the same direction as the elements in the regular receptive field. (Based on data from Tanaka et al., 1986).

Neurons in area MT can be distinguished from motion sensitive cells area V1 on the basis of their responses to four special types of motion stimuli: *apparent motion*, *plaid motion*, *transparent motion*, and motion *outside* the cells receptive field. MT cells can detect the apparent displacement of a flashing light (apparent motion) over much larger distances than V1 cells, a property that could be accounted for by the greater size of their receptive fields. This property may also help explain the fact that, in general, MT neurons are found to be sensitive to significantly higher stimulus velocities than V1 cells (Mikami et al., 1986b). In contrast, the temporal summation properties of MT cells may not be substantially greater than those of V1 cells, suggesting that spatial factors -- such as an increase in the separation of subunits in the cell's receptive field -- rather than temporal factors, determine the sensitivity of MT neurons to higher velocities of apparent motion. Finally, the range of preferences noted for MT neurons closely matches the range of human psychophysical thresholds for apparent motion detection (Newsome et al., 1986). Thus, MT neurons appear suitable for coding the direction and speed of apparent motion over a large portion of the visual field and a wide range of stimulus velocities. With respect to plaid stimuli, a substantial proportion of MT cells (approximately 20% according to one estimate; Movshon et al., 1989) are tuned to the perceived motion vector, rather than to the direction of any of the individual moving components, as shown in Figure 63, a property not found in area V1 (see Figure 46). MT cells are capable of integrating movement that occurs within their receptive field in different directions, and their firing correlates with the subjective perception of motion produced by the entire pattern.

Another domain that distinguishes MT cells from V1 neurons is their response to transparent motion. We saw earlier in this chapter that cells in area V1 display preference for a narrow range of motion directions. When stimulated with transparent motion stimuli, V1 cells increase their discharge rate when their preferred direction matches the direction of one of the moving components (see Figure 47). A given cell in area V1 can signal the presence of a single motion vector at each retinal location. In contrast, the majority of direction-selective cells in area MT demonstrate response suppression when stimulated with transparent motion stimuli, as illustrated in Figure 64. The amount of response suppression is typically smaller than the reduction in firing rate produced by an array of dots moving in the anti-preferred direction. The behavior of MT cells correlates with psychophysical data obtained from human observers. Thus, it has been shown that a moving surface is easier to detect when it contains a single motion vector, than when it consists of elements moving in two different directions. This outcome would be expected if area MT played an important role in the perception of motion, a position strongly favored by lesion data in primates (Schiller, 1993).

Although the sensitivity of MT cells to other, more complex forms of motion, such as rotation, expansion and deformation has not been studied extensively, it appears that individual cells respond to many types of motion without showing clear preference for one particular kind (Lagae et al., 1994). Such selectivity is first encountered in the medial superior temporal area, which will be discussed in the next section. Inhibitory processes have been implicated by a number of investigators in order to account for the direction selectivity of MT (Allman et al., 1985; Mikami et al., 1986a; Snowden et al., 1991). With respect to the responses of MT neurons to transparent motion several alternative mechanisms have been proposed. One mechanism may involve converging projections from V1 neurons that show opposite direction preferences.

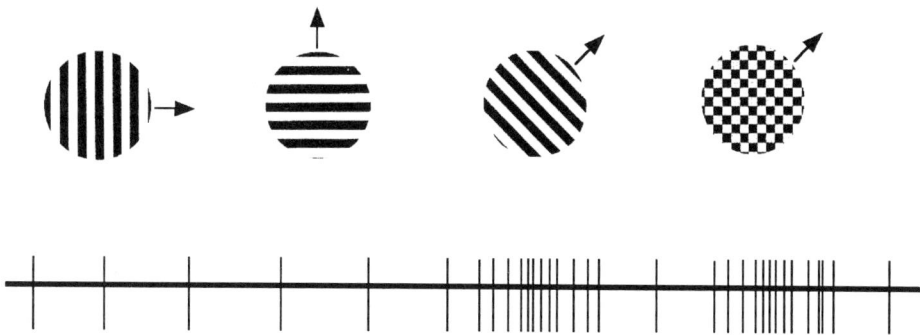

FIGURE 63.
Schematic drawing of the typical response of a MT cell to moving grating and plaid patterns. The cell's preferred direction of motion is first determined using sinusoid grating patterns. The cell also responds to the plaid pattern, i.e., to the perceived motion direction, although it does not show preference for either one of the two component gratings (left and left middle stimuli).

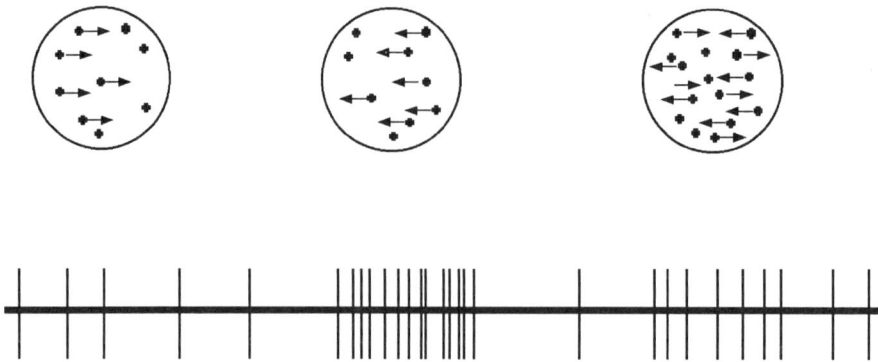

FIGURE 64.
Typical response of a MT cell to transparent motion. The neuron shows clear preference for left-ward motion (middle) but its firing rate decreases considerably when elements moving in the anti-preferred motion are added to the display.

It is possible that all of these inputs are excitatory and that the sign of the input (i.e. excitatory or inhibitory) from those V1 cells that signal motion in the antipreferred direction of the MT neuron becomes reversed via an intervening inhibitory interneuron (Andersen et al., 1993). A similar mechanism is likely to be involved in shaping the response profile of many MT cells associated with motion in the silent surround of their receptive field. These processes are represented in schematic form in Figure 65. An alternative mechanism may involve inhibitory interactions between several MT cells that have different preferred directions of motion.
The precise nature of inhibitory processes remains a matter of controversy (Mikami et al., 1986a; Snowden et al., 1991). One process may involve the linear summation of excitatory and inhibitory postsynaptic potentials that arrive on a given MT cell in close temporal synchrony. Thus, when motion in the anti-preferred direction is added to an array of elements that move in the preferred direction, the cell's output is reduced, as if the effects of the former stimulus are subtracted from the effects of the latter.

FIGURE 65.
Schematic representation of neuronal circuits likely to be responsible for direction selectivity in areas V1 and MT. Both inhibitory and excitatory processes may determine direction selectivity in V1 cells (although experimental data favor the latter mechanism). In this example, a dark bar moving from left to right first enters the receptive field of an OFF-center dLGN cell ("A") and then that of cell "B". Both P and M cells will increase their firing rate in close temporal synchrony, but M cells are more likely to produce a transient response, which is required for providing information regarding the relative timing of stimulation. The latter is important for coding the velocity of the moving stimulus. According to the facilitatory model, the excitatory inputs from cells "A" and "B" onto cortical cell "S" are integrated and the combined EPSPs are sufficient to excite the cell.

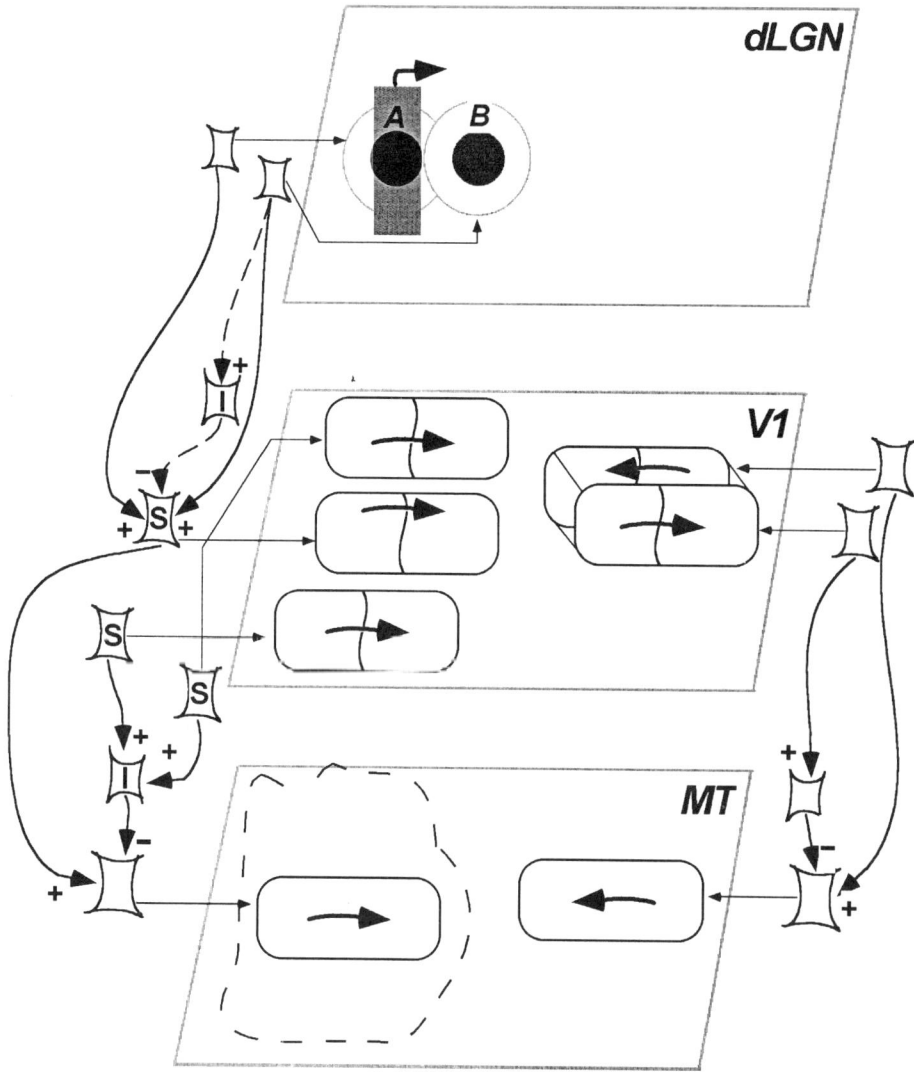

Summation of inputs does not occur when the receptive field of cell "B" is stimulated before that of cell "A", due to the particular spatial arrangement of the synaptic contacts within the dendritic arbor of cell "S". This model does not predict response suppression below baseline levels for cell "S" with stimulation in the antipreferred direction. The inhibitory model implicates inhibitory input from interneuron "I" that receives input from cell "B". When "B" is excited first, the interneuron is also excited and, in turn, either directly inhibits firing in cortical cell "S", or simply shunts away the excitatory potentials induced by subsequent stimulation of subregion A. The hypothetical circuit responsible for the response profile of MT to stimulation of the center and the silent surround of their receptive field is portrayed on the lower left-hand portion of the figure. On the right-hand side of the figure, a schematic circuit diagram, that is likely to be responsible for the responses of MT cells to transparent motion, is shown.

An alternative process may involve shunting inhibition (see Chapter 2), which could operate by reducing the likelihood of discharge in the postsynaptic cell in response to an independent excitatory input (Snowden et al., 1991).

If shunting inhibition were primarily active, one would predict that the cell's response to motion in the preferred direction would be reduced by a constant factor by the addition of motion in the antipreferred direction. This prediction was consistent with the results of further experiments conducted by Snowden and colleagues (1991). The most likely neurotransmitter for mediating the hypothetical process of shunting inhibition is GABA (Dreifuss et al., 1969).

The role of area MT in motion perception is supported by data from lesion experiments in the macaque. Chemical lesions (using ibotenic acid) that destroyed small portions of area MT were associated with a marked increase in the psychophysical threshold for the detection of certain kinds of visual motion (Andersen & Siegel, 1986, cited in Andersen, 1989; Schiller et al., 1990).The deficits were restricted to regions of the visual field corresponding to the portion of the retinal representation in area MT that was affected by the lesion. Other basic functions, such as contrast sensitivity, form perception, and stereopsis are largely unimpaired following restricted MT lesions (Schiller, 1993). Motion detection and even velocity discrimination for simple forms of visual motion are not entirely abolished following such lesions. It appears, therefore, that other visual areas can perform the neural operations required for certain aspects of visual motion analysis.

Area MT is part of the parietal visual pathway, which is presumed to be minimally involved in the analysis of visual cues based on wavelength. This view is supported by lesion studies in primates: lesioning this area does not typically cause any deficits in wavelength discrimination (Schiller, 1993). Nevertheless, many MT cells show sensitivity to wavelength: for instance many MT cells increase their firing rate in response to isoluminant moving contours, although they respond more vigorously to a moving stimulus that contains pure luminance contrasts (Saito et al., 1989). These findings do not necessarily contradict the view that area MT is not part of the brain mechanism responsible for color perception. It is a long-standing contention in human psychophysics that color and motion cues interact in the visual system. It is therefore likely that the sensitivity to chromatic transitions displayed by MT cells serves as a complementary process in the analysis of visual motion.

The behavior of these MT cells is one example of the fact that neuronal sensitivity to a particular sensory cue is often distributed across pathways or systems that perform different visual analysis functions. It is frequently the case that each system uses this information to signal the presence of a different visual attribute in a complex visual scene. In the previous example, the temporal pathway uses its sensitivity to wavelength variations to identify different colors, which can be used as cues for object identification, whereas the parietal pathway uses the same information to detect motion. This phenomenon is one example of *redundancy* in the nervous system. In this case, the redundancy, or overlap in the sensitivity to visual cues displayed by different cortical areas, is partial. In other words, the same type of visual input is represented in more than one area where it contributes to a *different* function.

The medial superior temporal area (MST)

Neurons with preferences for even more complex types of visual motion are encountered in another visual association region, the medial superior temporal area (MST). Two functionally distinct subregions have been identified within area MST. In the *ventro-lateral* region (MSTl), cells respond best to the motion of small stimuli, such as spots of light. This property is particularly suitable for guiding smooth pursuit eye movements. Small chemical lesions placed in this region result in severe difficulties in maintaining precise tracking of moving visual targets (more on the role of parietal areas in visual guidance in the next section). In the *dorsomedial* region (MSTd), cells show unique selectivity to different types of complex visual motion. This area will be discussed in more detail in the following section.

Sensitivity of MSTd cells to complex visual motion

MSTd neurons have very large receptive fields, which cover more than one quadrant and often extend into the ipsilateral visual field. In addition, receptive field size is relatively constant across a wide range of retinal eccentricities. These neurons show strong direction selectivity, respond best to large moving arrays, and are often selective for different types of complex motion, such as movement of an object in depth, *expansion* and *contraction* of the visual array, *rotation*, and *spiral motion*. These types of motion are considered as components of optic flow, the global visual motion generated by locomotion through the environment (Warren & Hannon, 1988) For instance, expansion corresponds to outward *optic flow*, and results from forward movement of the observed, whereas contraction involves inward optic flow, and results from backward movement. Expansion/contraction can also be simulated by radial motion of the elements of a large array, which appear to move either toward or away from a central point. During actual movement of the body, this point defines the axis of locomotion. In addition, simple translational motion (straight motion on the frontoparallel plane) is generated during eye movements and head rotation, while rotational motion results from head bending. Psychophysical experiments have shown that human observers can determine, with a high degree of accuracy, the direction of perceived self-motion on the basis of optic flow cues alone. Theoretically, decomposition of optic flow into the complex motion components listed above can provide accurate information about the direction and velocity of self-movement in relation to the layout of nearby objects.

Many MSTd neurons display selective responses for only one type of motion (rotation, expansion, contraction, translation, or spiral motion). The preference of some MSTd neurons for either rotation or translation (see Fig. 66) is consistent with psychophysical evidence that rotational movement can be perceived independently from simple translational movement (i.e., linear displacement). This suggests that the two types of motion are analyzed by separate neural mechanisms. Moreover, the preference of MSTd neurons for expansion or contraction on both dimensions (see Fig. 67) is in sharp contrast with the responses of MT cells, which show sensitivity to stimulus expansion or contraction along a single dimension, i.e., either along the width or along the length of a bar stimulus. Many MSTd neurons respond best to motion in depth under both monocular and binocular viewing conditions (Saito et al., 1986). Some cells respond selectively to a change in stimulus size (which corresponds to the visual effect produced as an object moves toward or away from the observer), whereas others prefer a change in interocular

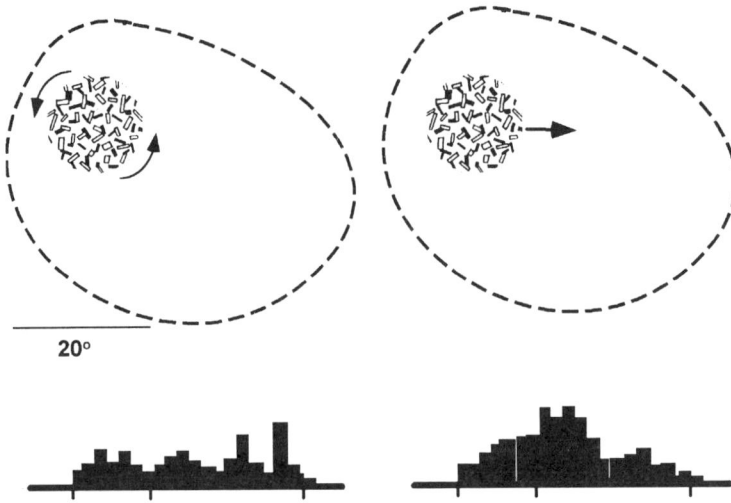

FIGURE 66.
Responses from an MSTd neuron that is selective for rotational motion. The cell responds vigorously to clockwise rotation and gives a very weak response to straight movement of the random element display. (Based on data from Saito et al., 1986.)

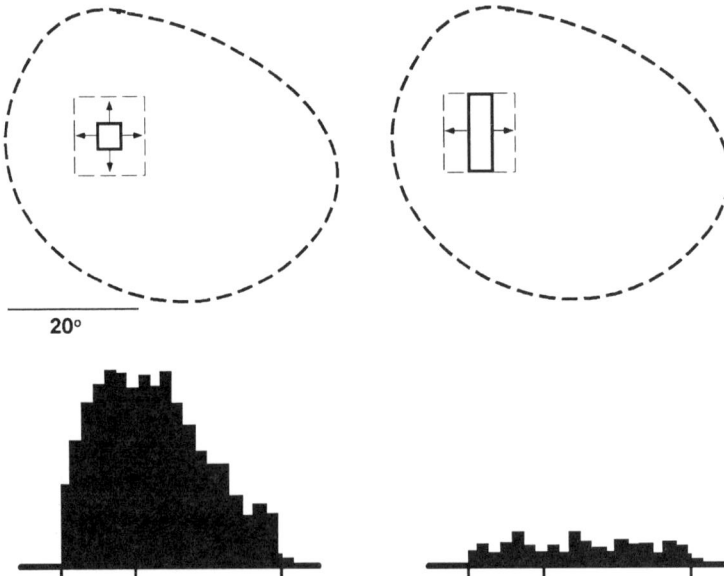

FIGURE 67.
Average discharge histograms from an MSTd neuron that shows preference for expanding motion (left). The cell responds to the expansion of the size of a circle but does not respond to expansion in only one dimension. (Based on data from Saito et al., 1986.)

disparity per se (Sakata et al., 1997). These cells appear sensitive to visual cues involved in coarse stereopsis and are suitable for signaling the presence of visual targets to be fix- ated with a change in vergence (i.e., by convergence or divergence of the lines of sight).

A substantial proportion of cells in MSTd respond to more than one type of motion, such as translation and expansion, rotation and translation and, occasionally, to all three types of motion. Duffy and Wurtz (1991a) termed the cells in the former class "double- component", and the cells in the latter class "triple-component" neurons.

The degree of selectivity for the direction of motion in a given neuron correlates with its sensitivity to different types of motion: "single-component" neurons showed the high- est selectivity, while "triple-component" neurons demonstrated the lowest selectivity among the three classes. Cells in the former class were those most likely to display tonic response suppression, whereas cells in the latter class were the least likely to display this phenomenon.

If a neuron is responding to the global motion pattern present in the visual array rather than to the motion of visual elements in certain parts of that array, it must show position invariance. Indeed, MSTd cells show the same selectivity for a particular motion pattern (such as the rotating textured surface shown in Fig. 68) regardless of its position within the receptive field. Note that in the example presented in Figure 68, local motion patterns within the same portion of the receptive field, like the region enclosed by the rectangle, change drastically when the stimulus changes position.

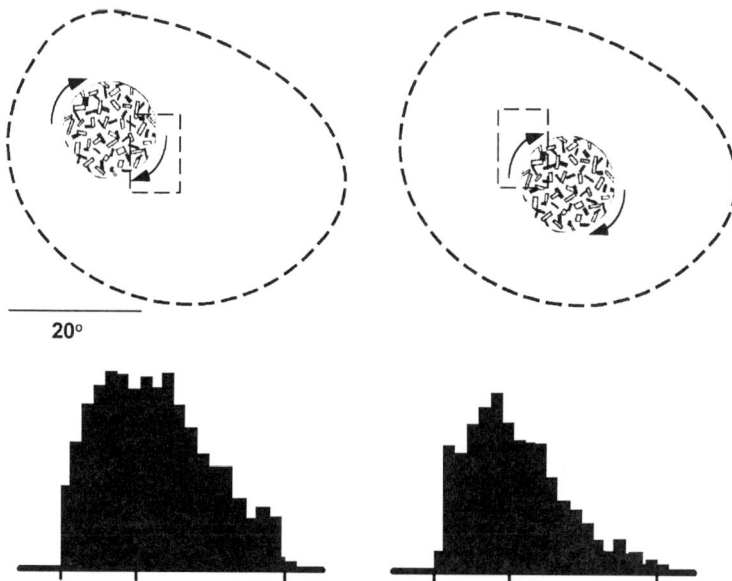

FIGURE 68.

The average discharge rate of an MSTd neuron does not change significantly as a function of the position of the stimulus (an array of rotating dots) in the cells receptive field. (Based on data from Saito et al., 1986.)

Neurons that show position invariance have also been found in area MT, but they can be easily distinguished from MSTd neurons as they have smaller receptive fields. Thus, while MSTd neurons can support position invariance almost *anywhere* within the visual field, MT cells cannot. Position invariance, however, renders individual cells unable to signal the precise location of a moving array in the visual field. For instance, the response of a cell selective to expansion could be used to signal forward movement of the observer, but the direction of movement could only be determined by pooling responses from a population of cells with similar sensitivity, but slightly offset receptive fields relative to each other (Lagae et al., 1994).

Another property that is a requirement for any mechanism that utilizes optic flow cues is the ability to integrate information regarding the movement of individual elements in the visual array over a large area of the visual field. This becomes possible for individual neurons in area MSTd due to the large size of their receptive field, as shown in Figure 69. Further, the mechanism must "smooth" velocity discontinuities that regularly occur in the optic flow array produced during locomotion. Thus, under natural conditions, elements in the center of the optic flow field move at a slower speed than elements in the periphery.

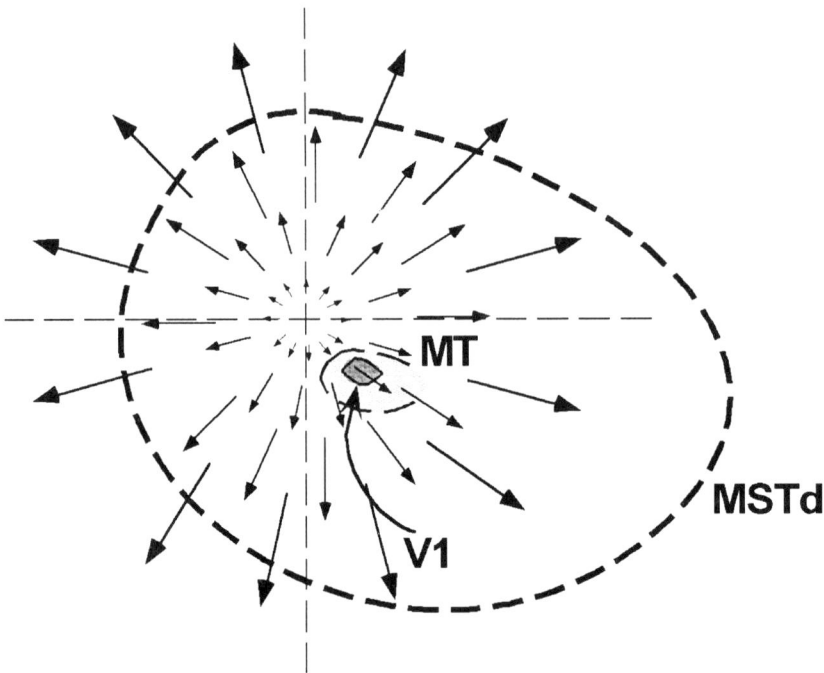

FIGURE 69.
Typical receptive field size for motion-sensitive neurons in areas V1, MT, and MST. The background represents an optic flow field, like the one produced by forward movement of the observer toward the intersection of the two dashed lines.

Efficient analysis of optic flow cues requires the engagement of a "smoothing" process capable of ignoring differences in local velocities. Indeed, neurons in area MSTd respond to a wide range of local velocities of the individual elements in these displays.

Moreover, the firing of some of these neurons is not significantly affected by visual cues that are not essential to the perception of optic flow per se, such as texture gradients, the density of visual elements, etc. (Duffy & Wurtz, 1991a). Thus, MSTd neurons appear to be primarily involved in analyzing the visual consequences of self-motion while "ignoring" visual movement that occurs regardless of the status of the observer. Their sensitivity to many different types of optic flow and their combinations makes them suitable for providing other brain systems with accurate visual information that could be used for the guidance of self-motion.

Two alternative mechanisms have been proposed to explain sensitivity for large-field motion in MSTd (Duffy & Wurtz, 1991b; Tanaka et al., 1989). According to one hypothesis, different subregions within the large receptive field of MSTd neurons display distinct preferences for motion direction.

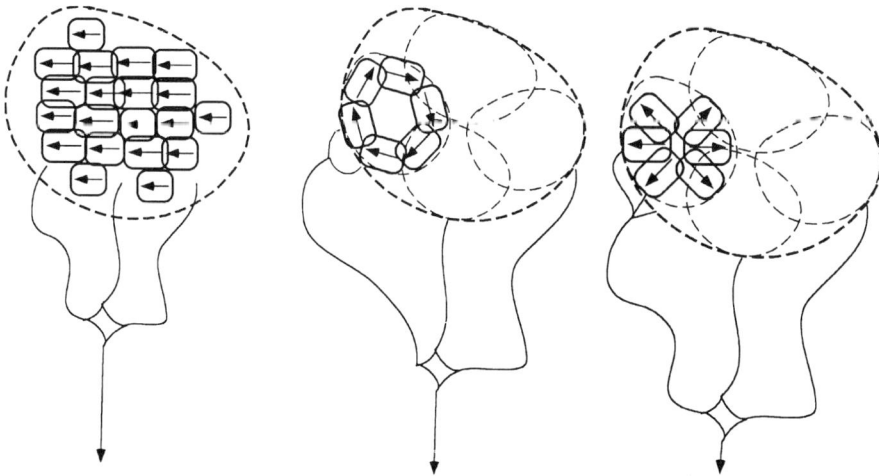

FIGURE 70.

Schematic drawing of the spatial layout of subregions in the receptive field of three MSTd cells that show selectivity to leftward translation (*left*), clockwise rotation (*middle*), and outward radial motion (*right*). Each subregion corresponds to the receptive field of at least one MT cell that shows selectivity for a particular direction of linear motion. The cell on the *right* receives input from MT cells, the receptive fields of which are arranged radially, whereas a circular receptive field layout is used to account for the receptive field of the MSTd neuron shown in the *middle*. Position invariance is explained by assuming that the receptive field of these MSTd neurons consists of several, spatially overlapping compartments (smaller dotted regions). Each compartment consists of similarly organized subregions that sample different portions of the visual field, and may reflect the integration of input onto a single dendritic branch. In this way, MST d cells can be driven by appropriate stimulation presented to any of its receptive field compartments. (Based on data from Saito et al., 1986).

In Figure 70, the receptive field of a hypothetical MST neuron is represented by the large rectangular area. Smaller rectangles represent functionally distinct receptive subregions. The selectivity of this neuron to a particular type of complex motion is determined by the spatial layout of the receptive field subregions.

According to an alternative explanation, each subregion is sensitive to the same type of complex motion that is found for large-field stimuli. In other words, sensitivity for a particular type of complex motion (expansion/contraction, radial movement, or rotation) is uniformly distributed throughout the receptive field. Duffy and Wurtz (1991b) devised several tasks for testing the two alternative mechanisms. A general conclusion that can be drawn from these experiments is that both types of mechanisms are probably present in MSTd neurons. The sensitivity of some cells appears to be determined by the former mechanism, whereas other cells respond in a manner compatible with the second mechanism.

In the human brain, only area MT has been localized using histological criteria (myelin staining, the monoclonal antibody CAT-301, and cytochrome oxidase staining). The MT-like area was found approximately 4-6 cm anterior to the occipital pole, in and posterior to the inferior temporal sulcus (Tootell & Taylor, 1995).

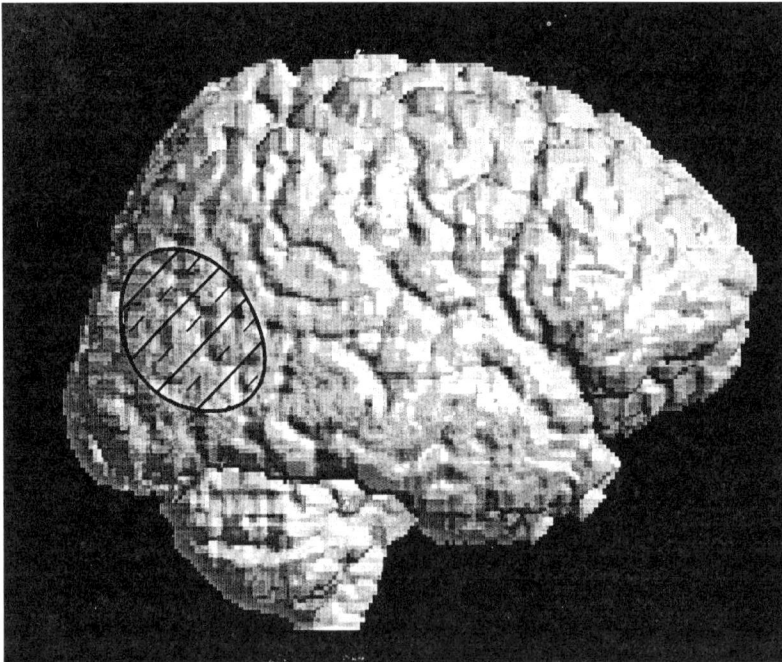

FIGURE 71.

The location of a lesion associated with a specific deficit in motion perception. The lesion, located in the vicinity of Brodmann's areas 19 and 39, is drawn on a model human brain. (Based on data from Zihl et al., 1991). Elevated levels of blood flow or population neuronal activity during stimulation with complex motion arrays have been found in the intact brain using non-invasive functional brain imaging methods.

This region is part of the cortex that shows evidence of elevated activity levels (in the form of evoked magnetic fields or measures of local blood flow and metabolism) in response to stimulation with complex motion (Demb et al., 1998; Vanni et al., 1997; Uutisalo et al., 1997). Although it is difficult to separate areas MT and MST on the basis of these reports, lesion studies also suggest that this region is indeed important for complex motion perception in humans. For instance, Zihl and colleagues (1991) studied the perceptual capabilities of a patient with bilateral lesions in the occipito-parietal junction, as shown in Figure 71.

This patient showed deficits that were restricted to the perception of moving stimuli. She had difficulty detecting rapidly moving spots of light and small displacements of patterned stimuli. Further, she was unable to perceive the apparent motion produced by flashing two adjacent spots of light in rapid succession. The specificity of her deficits was indicated by the fact that she displayed normal visual acuity, stereoacuity, and color perception. Interestingly, her ability to perceive spatial relations (such as the perception of the relative distance between two objects) was within normal limits. Her clinical picture showed many similarities with the pattern of perceptual deficits observed following chemical lesions in area MT in primates. However, it is too early to conclude that the disruption of motion perception described by Zihl and his associates was caused by damage to the cortical region that corresponds to the primate area MT or area MSTd.

Posterior parietal cortex

With the exception of area 7, posterior parietal cortex regions have received little attention until recently. Most of these areas lie along the banks and fundus of the intraparietal sulcus (see Fig. 21). There is consensus that posterior parietal regions are involved in visuospatial function and also that they provide crucial information for the guidance of eye, arm and head movements. However, the precise functional role of each area is often obscured by the diversity of stimulus attributes to which cells in these areas show sensitivity. Further, there is a scarcity of lesion studies in non-human primates, and such studies in humans are not directly comparable with primate electrophysiological investigations due to differences in testing procedures and tasks.

Stimulus attributes that posterior parietal neurons are predominantly sensitive to can be classified in three major categories: (1) those that convey information regarding the 3-dimensional shape of objects, (2) those that convey information regarding the spatial location of objects and the layout of surrounding space, in general, and (3) those related to visual motion. In addition, neurons in some parietal areas respond before or during saccadic eye movements. Neurophysiological response properties that pertain to each class of sensory attributes are described below.

Information regarding shape can be conveyed by visual input alone (through both monocular and stereoscopic cues), by somatosensory (tactile and proprioceptive) cues, or both. Neurons sensitive to visual, somatosensory input, or to both, have been found in several posterior parietal areas including the lateral intraparietal (LIP) (Sereno & Maunsell, 1998), the anterior intraparietal area (AIP), and the caudal portion of the intraparietal sulcus (c-IPS) (Sakata et. al., 1998). The prevailing view among researchers is that information on stimulus attributes that convey 3-D object features signaled by parietal cortex neurons is used to control hand posture for smooth object grasping and manipulation (Jeannerod et al., 1995). It is implicitly assumed that although the response properties of neurons along both the temporal and parietal pathways signal the 3-D shape of objects, this information is used for different purposes by each pathway, namely object identification in the former case and object manipulation in the latter. Sensitivity to visual motion is very common among posterior parietal neurons. Typically, areas that contain motion-sensitive neurons, such as areas VIP (Colby et al., 1995) and LIP (Eskandor & Assad, 1999), receive input from areas MT and MST. The degree of selectivity for motion direction and velocity found in parietal neurons clearly implicates these neuronal populations in the analysis of visual motion. It appears, however, that each parietal area contributes to different operations. For instance, the response profile of LIP neurons makes them suitable for signaling the trajectory of moving targets even under conditions of limited or intermittent viewing (Eskandor & Assad, 1999). This information could be used for guiding hand or eye movements toward moving targets. Many VIP neurons also signal the trajectory of a moving stimulus and often show preference for stimuli that move toward the animal's face (Colby et al., 1995). Interestingly, some of these neurons are multimodal, having a somatosensory receptive field aligned with the projected point of contact of the preferred visual stimulus. Information coded by these neurons may be used to guide head movements for capturing targets by mouth.

Posterior parietal neurons must also receive extraretinal input, signaling eye and head position and movement. From a computational point of view, this information is necessary for estimating the precise location of targets relative to either the direction of gaze or the center of the head, respectively. In turn, transforming retinal coordinates (i.e., the location of the target on the retina) into oculocentric or head-centered coordinates is important both for establishing an invariant map of spatial layout and also for planning accurate reaching movements toward preselected visual targets. This ability becomes crucial given that the precise mapping of the visual array on the retina changes constantly with movement of the head and eyes (often even *during* an ongoing reaching movement).

Sensitivity to spatial location regardless of the direction of gaze has been found in many cortical areas including MT, MST, area 7 and LIP. Recently, neurons that show narrow selectivity for particular head-centered locations have been found in area VIP (Duhamel et al., 1997). It appears, therefore, that several posterior parietal areas may contain head or body-centered representations of space, each specialized to support different functions. The relative importance of posterior parietal cortex for perceptual (visuospatial perception) and motor functions (guidance of eye and hand movements) is a matter of continuing debate.

NOTES

[1] According to recent estimates, only 5 to 20% of the total number of synapses in layer IV are made by geniculocortical afferents. The majority of the remaining synapses is formed by other neurons in area V1.

[2] Layer IV is also called *granular* or *afferent* layer.

[3] Axon terminals in the locus of the injection absorb this enzyme, which is then carried by the cells' retrograde transport system to the cell soma. In this way researchers can trace the course of an axon and identify its source and target areas.

[4] This method relies on the fact that the fluorescent agent is absorbed by the retinal ganglion cells and is transported transneuronally to the visual cortex. After the animal is sacrificed, the visual cortex is cut tangentially through layer IV and photographed to reveal the distribution of labeled neurons.

[5] A histological technique used for in vitro cell staining. Although it stains only a small proportion of cells in a given region, it has the advantage of revealing the entire structure of a stained neuron (including the dendrites, soma, and axon).

[6] Often referred to as isoluminant *color* contrasts, although as we noted in the introduction of this chapter, color (or hue) is the perceived property of, and therefore should not be identified with, *wavelength*.

[7] As noted in Chapter 3, every visual scene can be described as a composite of spatial variations of brightness and hue. Mathematically, each portion of a visual pattern can be decomposed into an finite number of sinusoid bar gratings of different spatial frequencies by means of the Fast Fourier Transform algorithm. The algorithm also specifies the strength of the contribution of each component frequency to the visual pattern in question. Theoretically, adding up each component, properly weighted, should reproduce the original visual image.

[8] The term "homonymous" refers to a defect that occurs in the same visual field in both eyes. Lesions located proximal to the optic chiasm always result in homonymous visual field defects. In contrast, lesions in both retinas, optic nerve, or chiasm cause various forms of heteronymous visual field defects, which could be restricted to one hemifield (in the former case – e.g., binocular scotoma), or extend to opposite halves of the visual field in both eyes (in the latter case – e.g., bitemporal hemianopia).

[9] This type of disparity cue is known as *positional disparity*. Other cells show receptive field disparities related to stimulus orientation (*orientation disparity*; von der Heydt et al., 1980) and the direction of movement (Cynader & Regan, 1978). We will restrict our discussion to positional disparity, which is by far the most well-studied and closely related to the perception of depth.

[10] Crossed disparity in a binocular image is created when the target is located between the eyes and the fixation point. Uncrossed disparities result when the target is located further away from the fixation point.

[11] This can be achieved experimentally by varying the relative proportion of wavelength bands (i.e., short, middle, and long) in the light used to illuminate the test surface. This is also a common phenomenon in everyday life. For instance, tree leaves continue to look green in the sunset when the sunlight contains a large proportion of long wavelengths and appears orange.

[12] Achromatopsia is a clinical condition in which the patient loses the ability to see color as a consequence of brain damage (Zeki, 1990). It should not be confused with *color agnosia*, in which the patient cannot recognize specific colors, but retains the conscious awareness of different colors, or with *color anomia*, in which colors can be perceived but the patient is unable to assign names to them. Some patients with achromatopsia retain the ability to detect isoluminant wavelength contours despite a complete lack of conscious perception of color. This phenomenon probably reflects the contribution of spared neocortical areas. These may include areas V1 and V2, as well as areas which are consid-

ered part of the parietal pathway. At least in primates, one of these areas (MT) contains cells that are sensitive to isoluminant color contrasts, although they do not show preference for particular wavelength bands, or any other indication of "color-selectivity" (Saito et al., 1989).

[13] Associative agnosia should be distinguished from *optic aphasia* (Lhermitte & Beauvois, 1973) which is characterized by impaired naming of visually presented items despite evidence that recognition per se may be preserved. Thus, the patient is usually able to categorize visually presented objects, gesture their use and meaning, and produce a name of a similar or semantically related object.

[14] These stages are virtually identical to Marr's (1982) computational model of object description.

[15] This notion bears an intriguing similarity with associationist trends that have prevailed in Philosophy since Aristotle, and were resurrected by the founders of the movement of (English) Positivism. In its current version, it incorporates knowledge regarding the sensitivity of neurons at different stages of the visual pathway to sensory cues and visual features. In only that sense, the modern version of associationism can be considered as an improvement over earlier versions.

[16] Perceived motion induced by a series of stationary lights, located at a small distance from each other, flashed in sequence at appropriate temporal intervals. The array is perceived as a moving spot of light.

Chapter 6

Conclusions

Continuity between cortical and subcortical visual pathways

The evidence summarized in the previous sections strongly supports the existence of functionally distinct pathways or systems in the extrastriate visual cortex that follow anatomically separate routes. Some authors have gone beyond this to propose that a direct correspondence exists between the retinocortical P and M pathways and the temporal and parietal visual systems, respectively (e.g., Kandel, 1991). This proposal is based on two major premises. First, the anatomical segregation of the P and the M geniculocortical afferents is maintained within the striate cortex (area V1). In addition, efferent cortico-cortical projections, from area V1 to a number of extrastriate areas follow anatomically separate routes. For instance, magnocellular input from V1 layer IVB is directed to area MT but not to area V4. This segregation is also maintained within the projections to a single cortical relay area (i.e., area V2). Second, the anatomical connections between the cortical pathways (e.g., Felleman & Van Essen, 1991) are not as extensive as the connections within a given pathway. Connections between pathways probably serve a modulatory function and do not carry the main flow of neural signals between the hierarchically organized cortical areas.

As already mentioned, there are many indications that the parvocellular and the magnocellular geniculocortical pathways remain anatomically segregated, at least to a certain extent, up to the primary visual cortex. To recapitulate, the axons that originate in the magnocellular dLGN layers project mainly to layer IVCa which is the primary source of afferents to layer IVB. On the other hand, parvocellular afferents terminate mainly in lamina IVCb which projects primarily to layer IIIB (to both blob and interblob compartments). However, both blobs and interblobs receive magnocellular inputs (Fitzpatrick et al., 1985). This observation is consistent with the finding that selective inactivation of the magnocellular dLGN layers (via direct application of GABA) can cause a substantial reduction of the visual responses of cells in the superficial layers of area V1 (Nealey & Maunsell, 1991). This evidence clearly indicates that that blob and interblob compartments receive inputs from the magnocellular pathway in addition to inputs from the parvocellular pathway.

According to a number of reports, the anatomical and functional separation between the parvocellular and the magnocellular pathway is, to some extent, maintained within area V2 (see Fig. 53). Lamina IVB is the primary source of the afferents from area V1 to the thick cytochrome oxidase-stained zones in area V2, whereas the projections from area V1 to the thin stripes and interstripes originate mainly in the superficial layers II and III

(Livingstone & Hubel, 1987). There is also evidence that neurons located in different cytochrome oxidase compartments in area V2 can be distinguished on the basis of their preferences for different sensory cues. For instance, direction selectivity and sensitivity to binocular disparity is more likely to be encountered among cells in lamina IVB (Hawken et al., 1988) and in the thick stripes in area V2 (DeYoe & Van Essen, 1985; Hubel & Livingstone, 1987). Further, it has been reported that V2 neurons that project to area MT are more likely to be found in the thick cytochrome oxidase stripes, whereas cells that project to area V4 are found in greater concentrations in the thin stripes and interstripes (DeYoe & Van Essen, 1985; Shipp & Zeki, 1985). In conclusion, several lines of evidence support the hypothesis that the projections from the primary visual cortex, to the histologically and functionally distinct compartments in area V2, originate in different laminae, which, in turn, contain different functional cell classes.

However, the evidence available at present is far from conclusive. Some of the previously mentioned reports are based on a very small number of observations. What's more, cytochrome oxidase labeling has its own limitations. For example, the cortical zones in area V2 that become stained by this enzyme, often cannot be readily classified as thick or thin stripes. In the macaque monkey, the species used in most studies, this problem is more severe because the pattern of staining is even less discernible (Hubel & Livingstone, 1987). Another limitation lies with the nature of single unit studies, which allow researchers to study only a minute sample of cells in a given cortical region. Finally, there are several indications in the literature that the anatomical separation of the ascending projections from area V1 to area V2 is certainly not complete. Given the methodological limitations mentioned previously, and the lack of conclusive data, researchers have a long way to go before they fully describe the relation between the parvocellular and magnocellular pathways, and the extrastriate neocortical visual systems.

The hypothesis that there is continuity between cortical and subcortical pathways makes two predictions with respect to the consequences of the selective destruction or reversible inactivation of a single retinocortical pathway. The latter can be achieved via chemical lesions confined to either the parvocellular or the magnocellular subdivisions of the dLGN. One prediction is that such a lesion should alter the response properties of cells in only those extrastriate areas that belong to the corresponding cortical pathway. For instance, the selective lesioning of the magnocellular pathway at the level of the dLGN should significantly reduce the stimulus-evoked responses of cells in area MT that belongs to the parietal pathway. Conversely, single unit responses in area V4 of the temporal pathway should be left unaffected. A second prediction that derives from this hypothesis is that damage restricted to a single retinocortical pathway would significantly impair behavioral performance on tasks that are critically dependent upon the corresponding extrastriate pathway. For instance, destruction of the magnocellular pathway should cause a significant impairment in the ability to discriminate complex motion, leaving shape, face, and color discrimination relatively intact.

In order to test the former prediction, Maunsell and his associates (1990) made small injections of either magnesium or lidocaine into the parvocellular or the magnocellular dLGN layers in order to induce selective, reversible inactivation of the P or the M pathway, respectively. At the same time, recordings were made from cells in area MT in the contralateral hemisphere in response to moving and flashing stimuli. Following injections into the magnocellular layers, stimulus-elicited activity was reduced in all MT cells that were tested.

In most cases, activity was completely suppressed. In contrast, blockade of the parvocellular pathway had minimal effects on the responses of MT cells. In a more recent experiment, either lidocaine or GABA was used to block neuronal transmission in selected dLGN layers (Fererra et al., 1994. It was reported that stimulus-evoked activity of cells in area V4 was as likely to become suppressed following injections into the magnocellular layers as it was following reversible inactivation of parvocellular layers. This suggested that neuronal activity in area MT depends primarily upon magnocellular geniculostriatal input, whereas both the magnocellular and the parvocellular pathway determine the responsiveness of most neurons in area V4. It is possible that the contribution of the magnocellular pathway to the temporal cortical system is mediated by magnocellular geniculocortical inputs transmitted via the supragranular layers in area V1 (primarily layer 4B; Nealey & Maunsell, 1991). Magnocellular input is then relayed to area V4 via areas V2 and V3 (Felleman et al., 1997).

According to the second prediction, if the magnocellular pathway is the primary source of afferent input to area MT, then its selective destruction should significantly affect performance in tasks that require operations for which area MT appears to be specialized, such as discrimination of the direction and the velocity of visual motion. To examine this question, Merigan and colleagues (1991) obtained thresholds for the discrimination of the direction and velocity of visual motion from two macaques before and after they had received unilateral chemical lesions in the magnocellular dLGN layers (see Table 2 for a summary of their results). The motion stimuli were always presented to one eye (on the side contralateral to the lesion), and within parts of the visual field that were represented in spared portions of the parvocellular layers (to ensure that the animal could still see them). Both animals were capable of performing the tasks after the lesions. Nonetheless, they showed elevated psychophysical thresholds for direction discrimination that was restricted to stimuli moving at high velocities, and also for velocity discrimination when the brightness contrast of the stimuli was reduced. In addition, the ability to detect drifting gratings was greatly impaired, but only at high temporal frequencies. Therefore, in every instance the animal was capable of performing the required function(s), suggesting that one should not speak of the magnocellular pathway as being uniquely specialized for any single complex visual function. Rather, the deficits displayed by the animals were due to reduced capacity of the geniculocortical afferent system to transmit neural signals that normally supply the parietal cortical system with information regarding the low spatial and high temporal frequency components of the visual array. These kinds of visual information are crucial for maintaining high levels of sensory (or psychophysical) sensitivity to rapidly moving visual images. The fact that magnocellular cells are capable of signaling rapidly changing visual events is consistent with the "transient" character of the responses of magnocellular neurons. The reported low sensitivity of the latter to high spatial frequencies can be accounted for by their lower sampling density compared to the parvocellular neurons (Blakemore & Vital-Durand, 1986; Crook et al., 1988). The finding that the magnocellular geniculocortical pathway is the primary source of input to the parietal cortical system is consistent with the fact that the former is capable of transmitting the kind of information that is critical for the visual

Table 2. Deficits produced by selective lesions to the parvocellular
or magnocellular layers of the dLGN.

Function		Parvocellular	Magnocellular
Stimulus discrimination based on hue		Severe	--
Detection of isoluminant contrast		Mild	--
Pattern discrimination	High SF	Severe	--
	Low SF	Moderate	
Texture discrimination	High SF	Severe	--
	Low SF	Moderate	
Shape discrimination	High SF	Severe	--
	Low SF	Moderate	
Brightness discrimination	Low SF	--	--
Contrast sensitivity	High SF	Severe	--
	Low SF	Mild	Mild
Stereopsis	Fine	Severe	--
	Coarse	Severe	--
Flicker detection	Low contrast	--	Severe
	High contrast	--	Mild
Motion detection	Low contrast	--	Severe
	High contrast		Mild

Based on data from Merigan (1989); Merigan et al. (1991); Schiller (1993), Schiller et al. (1990, 1991). Dashes indicate no deficit. Abbreviations; SF: spatial frequency.

analysis functions that the parietal system is specialized to perform (Merigan & Maunsell, 1993).

The results of selective lesions to the parvocellular layers of the dLGN have been examined by Schiller and Logothetis in a series of studies, the result of which are also summarized in Table 2.

The role of visual association cortices in the brain mechanism responsible for the analysis of visual cues

Several lines of evidence strongly suggest that visual association areas in the primate brain are highly specialized regions that form two rather distinct anatomical pathways or systems. Early reports indicate that receptive field properties in certain areas are stimulus-specific and that neurons that show selectivity for a particular sensory cue (such as wavelength) are concentrated in certain regions (e.g., area V4) and not in others. Neurons that respond selectively to different sensory cues, such as one or more of the parameters of visual motion, are found in great numbers in other areas (i.e., MT and MST).

A crucial issue to be addressed is how neural signals that convey information on a variety of visual attributes become integrated to produce a unified percept. A relevant finding here is that preference, at the cellular level, for a particular visual attribute is rarely restricted within a single visual area. For instance, a recent study reported that area V4 contains cells that are sensitive to the parameters of visual motion. Moreover, the sharpness of tuning to one parameter of visual motion (velocity) was not different between the cells in areas V4 and MT (Cheng et al., 1994; see also Baylis et al., 1987). Thus, sensitivity to visual motion is not an exclusive property of cells in the parietal pathway. This is consistent with the finding that the responsiveness of cells in both area V4 and area MT depends upon the common input provided by the magnocellular pathway (Ferrera & Maunsell, 1991). It appears, however, that V4 is not indispensable for the ability to perceive motion, as indicated by the effects of experimental lesions summarized in Table 3. How the sensitivity to motion of individual units in area V4 actually contributes to perception is unclear at present. Conversely, neurons that respond to isoluminant color contrasts have been found in area MT (Saito et al., 1989), although ablation of this area does not lead to any detectable deficits in the ability of the animal to make color discriminations. The fact that cells capable of analyzing visual motion exist in an area traditionally considered as part of a visual system exclusively specialized for the analysis of pattern and color, is sufficient to cast doubt on the long-held view that a sharp distinction exists between the parietal and the temporal pathways.

Another conclusion that has emerged after three decades of lesion studies on the visual cortex is the refutation of the notion that a single visual area is exclusively responsible for the operations that lead to the analysis of a particular sensory cue (i.e., contrast, orientation, etc.) or to the perception of a given stimulus attribute (i.e., shape or depth). Thus the complete ablation of one or more association areas is typically *not* associated with severe deficits when the animals are tested with simple stimuli. For instance, visual acuity and wavelength discrimination is not severely affected following damage confined in area V4 (Heywood & Cowey, 1987; see also Table 3). Similarly, lesions in area MT do not, as a rule, impair the ability of the animal to detect moving stimuli. Lesioned animals showed clear deficits only in tasks where performance required the ability to perceive a certain stimulus attribute in a complex stimulus. For instance, lesions in area V4 impair the ability to discriminate hue when the wavelength composition of reflected light is variable (i.e., under conditions that require color constancy; Wild et al., 1985). Therefore, although area V4 may not be responsible for the entire set of operations that lead to color perception, its integrity appears to be necessary for perceiving color under certain

Table 3. Deficits produced by selective lesions to area V4 or MT.

Function		V4	MT
Stimulus discrimination based on hue		Moderate	--
Pattern discrimination	High SF	Moderate	--
	Low SF	Mild	
Texture discrimination	High SF	Moderate	--
	Low SF	Mild	
Shape discrimination	High SF	Moderate	--
	Low SF	Mild	
Brightness discrimination	Low SF	--	--
Contrast sensitivity	High SF	Mild	Mild
	Low SF	--	Mild
Stereopsis	Fine	Mild	--
	Coarse	--	--
Flicker detection	Low contrast	--	Severe
	High contrast	--	Moderate
Motion detection	Low contrast	--	Severe
	High contrast	--	Moderate

Based on data from Heywood & Cowey (1987), Schiller et al. (1990, 1991). Dashes indicate no deficit. Abbreviations, SF: spatial frequency.

conditions. In general, however, it appears that the organization of the extrastriate cortex is characterized by a great amount of redundancy, both within, as well as between, the putative visual systems.

A final issue that emerged in the previous chapter concerned the manner with which sensory cues are signaled by the activity of neural units. Early reports of individual cells that responded to a particular stimulus (i.e., the face of a given individual) with remarkable specificity led investigators to hypothesize that the activity of individual cells was sufficient to signal the presence of a behaviorally relevant stimulus in the visual environment (*sparse coding theory,* also referred to as the *neuron doctrine*).

However, subsequent studies have shown that neurons in visual association cortices (i.e., the inferior temporal cortex) respond in a graded fashion to a wide range of stimuli. Further, it was found that neighboring cells have the tendency to show overlapping, yet not identical, stimulus preferences. Cells that show sensitivity to a particular set of stimulus characteristics tend to cluster together in cortical patches that bear distinct similarities with the columns in the primary visual cortex. Moreover, multielectrode recordings reveal population response profiles that vary according to certain stimulus characteristics. These pooled neuronal responses may be sufficient to signal the presence of such complex stimuli as the face of a particular individual with a high degree of accuracy. These findings provide support to the notion that complex visual stimuli are coded in the activity of subpopulations of neurons. Neuronal populations are capable of developing sensitivity to various combinations of visual (and possibly also multimodal) features through repeated exposure. These neuronal properties meet the requirements of recent neuropsychological models of semantic representation. These models postulate distributed coding and hypothesize the existence of cortical areas in which sensory-cue selective responses of individual neurons may be combined in order to encode the complex features of real-life objects.

Chapter 7

Why Study Development

Precocious species like humans and primates are born with perceptual capacities that are not fully developed. For instance, visual acuity is on the average 60 times lower in a healthy human newborn than in the typical adult observer (Banks & Dannemiller, 1987). Other visual capabilities in the human infant do not emerge before the first four to six months of life, such as the ability to perceive depth from binocular stereoscopic cues. These and other similar observations have triggered several questions regarding the relation between perceptual and neural development, two of which will be examined in some detail in the following chapters. The first question concerns the degree to which the sensitivity of the developing nervous system is actually reflected in the perceptual capacities of young infants, while the second concerns the relative contribution of environmental and genetic factors to the development of the brain mechanisms responsible for perceptual functions.

With respect to the former issue, it is possible that the limits of perceptual sensitivity are determined exclusively by the level of nervous system maturation. Conversely, it is conceivable that perceptual development lags behind the maturation of the brain mechanisms that support the corresponding perceptual functions. Such discrepancies might be due to peripheral factors (i.e., maturity of receptor organs), to neural yet non-sensory factors, or to both. Non-sensory factors may include the ability to organize and execute the behaviors monitored by researchers to assess the infants' perceptual abilities, and the motivation to respond appropriately in a particular experimental setting. It is, therefore, important to have a clear picture of the developmental course of all the components, peripheral and central, that mediate sensation. Finding answers to this basic question is expected to have two important implications. First, and foremost, it will help determine the limitations imposed by the maturational state of the nervous system upon perceptual capabilities at each stage of development. For instance, it is known that the level of development of the human retina at birth permits a maximal spatial resolution of approximately 20 cycles/degree. Even if all the other components of the visual system, that normally contribute to spatial resolution, were fully mature, the immaturity of the retina would still limit visual acuity to a level that is at least one order of magnitude lower than the acuity of the average adult observer (Banks & Bennett, 1988). Second, by establishing the course of normal development, and identifying the processes involved in it, researchers will accomplish a major step toward understanding the pathophysiology of clinical conditions that interfere with normal sensory development. Such conditions include congenital abnormalities of the ocular media (such as *catarax*), severe imbalance in the refractory power of the two eyes (*anisometropia*), and misalignment of the two lines of sight (*strabismus*).

The second question, concerning the role of genetic factors in the development of perceptual functions, acquired new impetus after the discovery of the role of the DNA molecule (Crick & Watson, 1954) which substantiated the notion that the phenotypic characteristics of living organisms are genetically determined. In the era that followed this discovery, researchers attempted to establish how, and to what extent, a given phenotype is determined by the genome. Without doubt, the most problematic issue in this line of research concerns the nature of the involvement of the genetic code in the development of specific phenotypes. From the beginning of this endeavor it became evident that the genetic code could not contain detailed instructions for guiding the growth of each and every individual neuron. This led to revisions of the original genetic view, according to which the genetic code specifies a general *program* for the development of the nervous system. Attempts to define the processes which are involved in the storage and execution of such a program have created a great deal of confusion. It is now generally accepted that a successful genetic program should be capable of providing a satisfactory explanation for the growth of numerous components that interact with each other and are also affected by environmental events. But how could a genetic program "anticipate" not only the growth of other components of the sensory system but also the characteristics of the environment? Certain aspects of development, such as the synthesis of proteins from amino acids, cell division, migration, and neuronal differentiation, appear to have a strong programmatic component. In contrast, strict genetic programming is not likely to determine the architecture of sensory systems.

Diametrically opposite to this approach are various empiricist models proposing that the growth of neuronal connections and the development of functional properties of neural systems depend heavily upon environmental factors. In order to account for the findings that certain components and functional properties of the neuronal machinery emerge before the onset of sensory experience, several revisions to these models were proposed. A prevalent notion in these models considers development as an *epigenetic* process, rather than as the expression of a rigid genetic code. The advantage of epigenetic theories is that they take into account the role of the organism as an active receiver who can alter the nature of external input (Held & Hein, 1964).

At the cellular level, two key concepts that were developed in order to reconcile the strict genetic and empiricist views are: 1) *selective stabilization* of synapses and, 2) *modification of synaptic efficacy* that depends upon neuronal activity. Selective stabilization presupposes that there is an initial overproduction of neurons and synapses in the early stages of development. When neuronal signaling begins, a subset of these synapses becomes stabilized, while other synapses, afferent terminals, and even neurons, degenerate. Currently available data indicate that, especially in primates, selective stabilization involves primarily the rearrangement of synaptic terminals across target cells and, to a lesser extent, the elimination of neurons or neuronal processes. Synaptic modification is a generic term used to describe activity-dependent changes in the efficacy of synapses. This phenomenon often depends upon the temporal correlation between pre- and postsynaptic firing.

Three decades of intensive research in the field of developmental neuroscience have established the notion that the maturation of the brain mechanisms for vision depends upon visual experience. Alterations in the quality and/or the amount of visual experience during development can have irreversible effects on both neuronal and perceptual maturation. For instance, partial occlusion of the lens of the eye by congenital catarax typically prevents the development of normal spatial resolution, if it occurs during the first year of life in humans (Jacobson et al., 1982). Further, congenital misalignment of the two eyes can cause a permanent deficit in visual acuity and stereoscopic depth perception, if left uncorrected in the first one or two years of life (von Noorden, 1981; Banks, Aslin, & Letson, 1975).

A large body of research, starting with the pioneering work of Wiesel and Hubel (e.g., Wiesel & Hubel, 1970), has shown that uninterrupted vision through both eyes is necessary for the normal development of certain aspects of the fine anatomy and physiology of the visual system. For instance, occlusion of one eye during the first few months of life in the macaque monkey prevents the formation of ocular dominance columns in layer IVC of the primary visual cortex (LeVay et al., 1980). Disruption of normal binocular vision, that lasts as little as one month in infant primates, prevents the emergence of binocular neurons in the non-afferent layers of area V1 (Crawford et al., 1983). These and other similar phenomena suggest that the structural and functional organization of the nervous system is highly susceptible to changes in the characteristics of proximal stimulation during development. The ability of the nervous system to undergo long-lasting changes in its structure and function associated with environmental constraints is known as *neural plasticity*.

In the following chapters six special topics are examined that are relevant to the general issue of genetics versus environment. The first topic concerns the precise *timing* of the various phenomena of plasticity. Researchers quickly realized that the degree of susceptibility of certain aspects of nervous system organization is greater during certain periods of development and minimal in the adult organism. The fact that alterations in normal visual experience are capable of changing the course of perceptual and neural maturation, even if they are imposed for very brief periods during a particular age window, suggests that sensory input is more crucial during certain periods of development and less important during others. These and other similar observations led to the concept of the *sensitive* or *critical period*. Research on the timing of plasticity phenomena addresses two basic questions: 1) what is the earliest age at which the normal development of a particular phenotype (such as visual acuity) becomes susceptible to afferent input and sensory experience and, 2) at what age a particular phenotype becomes immune to drastic alterations in visual input.

The second topic deals specifically with the impact of visual input on phenotypic development. Afferent input may be involved in phenotypic maturation in one of the following ways, which are presented in schematic form in Figure 72. The first is *facilitation*. This implies that the phenotype is present, in rudimentary form, before the onset of external input, and visual input is required to ensure further growth. The emergence of ocular dominance columns in the primate visual cortex before the onset of visual experience, and their subsequent refinement under the influence of such experience, is one example of facilitation (Rakic, 1991). An alternative mode of environmental contribution to development is *maintenance*. This refers to the situation in which a particular type of

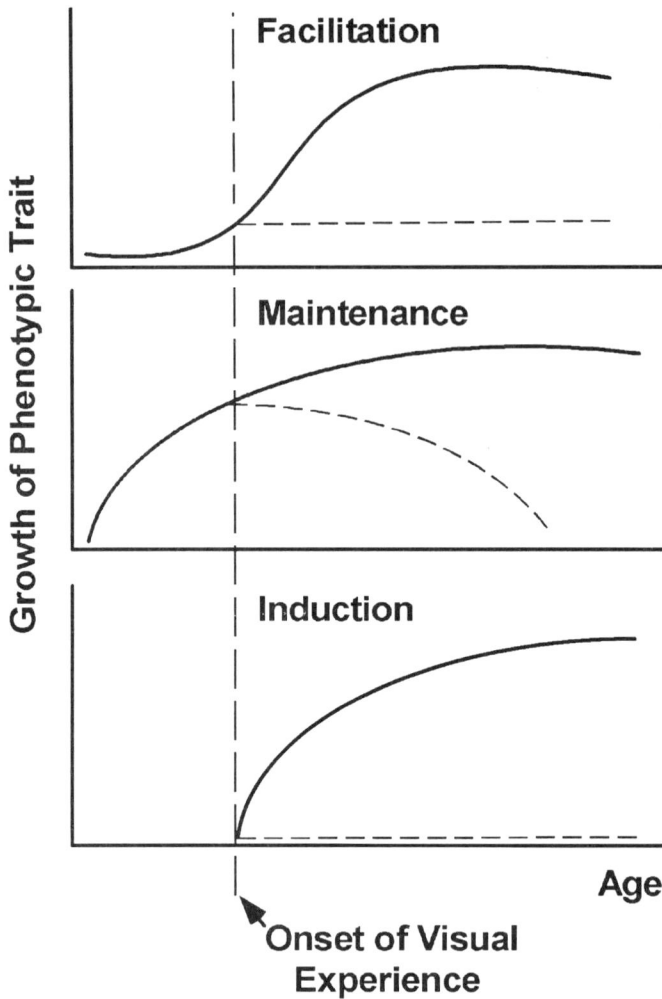

FIGURE 72.
Schematic rendering of the contribution of visual experience on the emergence and subsequent growth of a phenotypic trait (neuronal property of perceptual capability). Solid curves represent normal growth and dashed lines growth in the absence of appropriate visual input.

sensory input is required in order to preserve one or more neural properties that have emerged before the onset of sensory experience. To some extent, the effects of sustained exposure to visual contours of a particular orientation on the distribution of orientation preferences in the primary visual cortex can be considered as an example of the phenomenon of maintenance. *Induction*, theoretically, is the third alternative process through which visual input can affect neuronal and, consequently, sensory maturation. In this case, the initial emergence of a particular neuronal property clearly occurs after the onset of relevant visual experience. This process appears to be involved in the development of stereoscopic depth perception in humans, a phenotype that emerges several months after birth (Fox et al., 1979).

A third topic that has attracted much research attention is the extent to which a particular perceptual or neural phenotype depends upon specific components of visual input. For instance, it appears that normal development of spatial resolution requires stimulation with patterned visual contours and not simply stimulation with diffuse light. The processes involved in such phenomena likely operate in the visual cortex. In addition, one or more "central" processes may be involved in regulating the effects of visual input.

A fourth topic is whether plasticity in the developing organism is not only quantitatively but also qualitatively different from the phenomena of plasticity found in adult individuals of several species. One of the most well documented examples of mature or *adaptive plasticity* is the synaptic reorganization that takes place in the modality-specific neocortical areas following a focal lesion, deafferentation, or even persistent peripheral stimulation.

A fifth topic that has received extensive attention is whether different phenotypic traits have different critical periods. For instance, it appears that the critical period for synaptic stabilization takes place earlier in layer IV of the primary visual cortex of the cat than in the non-afferent layers.

The sixth and final topic concerns the neural mechanisms that underlie developmental plasticity. At the macroscopic level two general processes have been implicated in the development of the brain mechanisms responsible for vision, namely *cooperation* and *competition* between afferent fibers. At the cellular (microscopic) level it is believed that these processes are linked to activity-dependent mechanisms involved in synaptic stabilization.

In general, development is characterized by both *constructive* and *regressive* phenomena. The former include the production of neurons, neuronal processes (dendrites and axons), and synapses. Regressive phenomena include the partial elimination of neurons, axons, and synapses, and occur primarily after birth. The neurons that will later form a particular structure undergo a series of mitotic divisions in a remote region of the embryonic brain. Subsequently, they migrate toward their final position toward the developing brain structure. Soon after their migration, neurons send axons that form synapses with other cells in either the same area (e.g., intrinsic connections in the primary visual cortex), or in a different, often remote, area (e.g., the synapses formed by the geniculocortical axons in the visual cortex). As we will see in more detail below, pathfinding during neuronal migration and, in most cases, the guidance of outgrowing axons, do not depend upon the presence of electrical activity in the neurons involved. The early stages of development are characterized by an overproduction of new cells, dendritic branches, axonal collaterals, and synapses. Subsequently, there is a marked reduction in the number of cells, dendritic and axonal processes and, naturally, an elimination of a considerable number of the already formed synaptic connections. The timing of each of those processes varies widely between species and across different brain regions in a given species. Elimination phenomena appear to be selective and depend in most instances upon the amount, efficacy, and specificity of neuronal signaling among neighboring neurons.

Macaque (weeks)

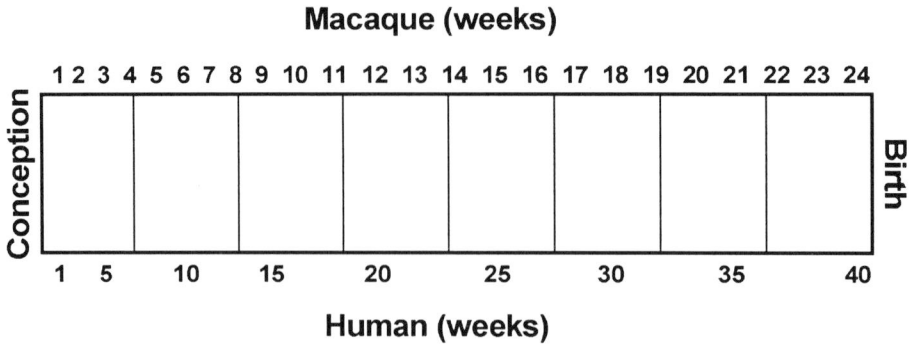

FIGURE 73.
Conversion chart for the gestational period in humans and macaque monkeys.

Many methodological approaches to the study of perceptual and neural maturation and plasticity, involve experimentally induced alterations in the organism's sensory environment.

For instance, one of the most widely used techniques is *visual deprivation*, which involves partial or complete blockade of visual input to one, or to both eyes simultaneously. Naturally, none of these techniques can be used with human infants due to their invasiveness. This created a need for an animal model of both the mature and developing visual system. Early studies on visual developmental plasticity were conducted on kittens. Despite the significant amount of information generated by these studies regarding the general principles of neural maturation, the extensive differences that exist between the human and the feline visual system rendered the cat model less than optimal. Researchers soon turned to non-human primate species in search of a better model, taking advantage of the fact that the human visual system is very similar to the visual system of Old World monkeys (like the macaque). In addition, visual capabilities, as measured by psychophysical methods, are very similar across species. At present there is a sufficient amount of normative data concerning the development of the visual system in humans and primates to justify the generalization of conclusions drawn from primate experiments to the human visual system. The age conversion chart presented in Figure 73 may be used to simplify across species comparisons.

For instance, the quality of the ocular media appears to be quite adequate at birth in both species (see Boothe et al., 1985 for a comprehensive review). In humans, the size of the eyeball undergoes a relatively smaller change (two- to three- fold) compared to a 20-fold increase in body size. The distance between the cornea and the retinal plane (axial length) increases by about 8 mm (which corresponds to a 50% increase) between infancy and adulthood. A similar increase is reported in three primate species (Blakemore & Vital-Durand, 1986a). With respect to refractive[1] accuracy, both human and primate infants show signs of hyperopia soon after birth and achieve normal refraction (emmetropia) in late infancy. A higher incidence of astigmatism reported in both species in infancy. The course of improvement in accommodation accuracy in the human infant parallels that of certain primates if age in thelatter species is multiplied by a factor of four. Specifically, accommodation slopes reach near-adult levels at approximately four months of age in the human infant (Banks, 1980) and at five weeks in primates. Optokinetic nystagmus[2] can be elicited in both species soon after birth. Another striking similarity is that in both human and primate infants younger than about two months or two weeks, respectively; monocular optokinetic nystagmus can only be elicited in the temporal-to-nasal direction. On the average, neonates of both species show approximately equal grating acuity (0.5 to 1 cy-cles/degree). Acuity improves about four times faster in primates in order to reach adult levels (30 to 60 cycles/degree) by the end of the first year. These similarities in the course of early visual development among humans and nonhuman primates, add validity to attempts to generalize, to the human infant, experimental data obtained from primate studies on the maturation of the visual system.

The remaining chapters of this book describe the major milestones in the development of the structural characteristics of the visual system in humans and primates. The focus is on four structures that have been studied extensively: the retina, dorsal lateral geniculate nucleus, optic tectum, and primary visual cortex. Whenever data from primates are not available, findings from other mammals (such as the cat and the ferret) and non-mammalian species (such as the goldfish and the frog) that have served as models of certain developmental processes are reported. The discussion then focuses on the development of key functional properties of visual neurons, such as spatial resolution, orientation selectivity, eye preference, binocularity, and sensitivity to retinal disparity. An attempt is made to relate these properties to the growth of corresponding sensory capabilities. Finally, some of the processes that appear to be involved in the normal development of these properties, as well in the phenomena of developmental susceptibility and plasticity, are examined.

NOTES

[1] Refraction refers to the "bending" of light as it passes through the optical components of the eye. The operation that changes the amount of refraction in order to compensate for variations in the distance of fixated objects from the viewer is called *accommodation*. The most important component of the accommodation system is the lens, the shape of which is controlled by the ciliary muscle. Relaxation of the ciliary muscle flattens the lens and allows objects located at a greater distance from the point of initial fixation to form a sharp image on the retina. Conversely, contraction of the ciliary muscle causes an increase in the thickness of the lens allowing objects located closer than the point of fixation to form a sharp retinal image. *Refractive error* refers to the amount of mismatch between the plane of the retina and the plane onto which visual images are actually projected. Excessive refractive power causes objects to be projected in front of the retina, whereas insufficient refractive power is associated with images being projected "behind" the plane of the retina. In both cases the image of the object appears blurred (defocused). When refractive error occurs systematically due to an abnormality in the optic media we speak of *hyperopia* in the former case and *myopia* in the latter case. In *astigmatism* the refractive surfaces of the eye have different curvatures and this results in different kinds and/or amounts of refractive error depending upon the orientation of the stimulus.

[2] Optokinetic nystagmus (OKN) is manifested as a series of rhythmic eye movements elicited when the subject looks at a pattern (usually a black and white grating) that moves continuously in either the horizontal or vertical direction. OKN consists of two phases: a slow phase of pursuit-like movements in the same direction as the moving visual stimulus, and a fast phase comprised of a saccadic movement that brings the eye back to its original position. The optokinetic reflex helps stabilize retinal images in an environment of constantly moving visual stimuli.

Chapter 8

Maturation of the Retinocortical Pathways

Development of the eye and the retina

The optical characteristics of the eye change little between birth and adulthood. Axial length is approximately 60% of the corresponding distance in the adult, a difference that could lead to reduced retinal image size in the eye of the neonate. However, the reduced axial length is compensated by the smaller size of the pupil. Thus, the amount of light that falls on the retina per squared degree is essentially adult-like at birth (Banks & Bennett, 1988). In the retina, ganglion cells mature earlier than any other type of cell in both humans and primates (La Vail et al., 1992). Soon after the final stage of their development they begin to send axons to the dLGN (Hendrickson & Rakic, 1976). The initial overproduction of retinal ganglion cells is followed by a sharp reduction in the number of optic nerve axons that occurs around embryonic week 14 in primates and between the 20th and 30th embryonic week in the human fetus (Provis et al., 1985). Some researchers believe that the elimination of ganglion cell axons is selective and largely restricted to those axons that synapse in the inappropriate eye-specific layer of the dLGN. This proposal is consistent with the observation that the emergence of eye-specific laminae in the dLGN coincides with the peak of ganglion cell death. In humans, the myelination of optic nerve fibers is nearly complete by two years of age (Magoon & Robb, 1981).

The regional maturation of the retina follows a rather peculiar course. With respect to the timing of cell production and synaptogenesis, the central region of the retina appears to develop earlier than the peripheral regions (La Vail et al., 1992). In contrast, the structure of rod receptors and photoreceptor spacing in the peripheral retina are quite mature at birth in both humans and primates. However, with regard to the properties of the fovea that are most closely related to its ability to transduce visual images with a high degree of resolution, dramatic changes take place during infancy in primates and in humans. For instance, the diameter of the rod-free foveola shrinks to about half its size in the newborn human infant. Shortly before birth retinal ganglion cells and other supporting cells that initially extend into, and cover the central retina progressively move laterally so that in the first postnatal months the incoming light can reach the central fovea unobstructed. At the same time a marked increase in the packing density of foveal cones is noted (Yuodelis & Hendrickson, 1986). Between birth and adulthood, average cone spacing decreases by a factor of two (Banks & Bennett, 1988). Presumably, this change is due to a migration of cones from the fringes of the foveal depression into the central fovea. Microscopic examination of cone shape at birth reveals that the inner segment, which is broader and shorter in infancy, does not fully mature before the age of 45 months. As shown in Figure 74, the diameter of the outer segment remains relatively unchanged while its length undergoes a 10-fold increase that continues beyond this age (Yuodelis & Hendrickson, 1986). Maturational changes related to the structure and the distribution of photoreceptors is of primary importance for the development of spatial resolution in the first few months of life. A more detailed discussion of the factors involved in this process is deferred for a later section.

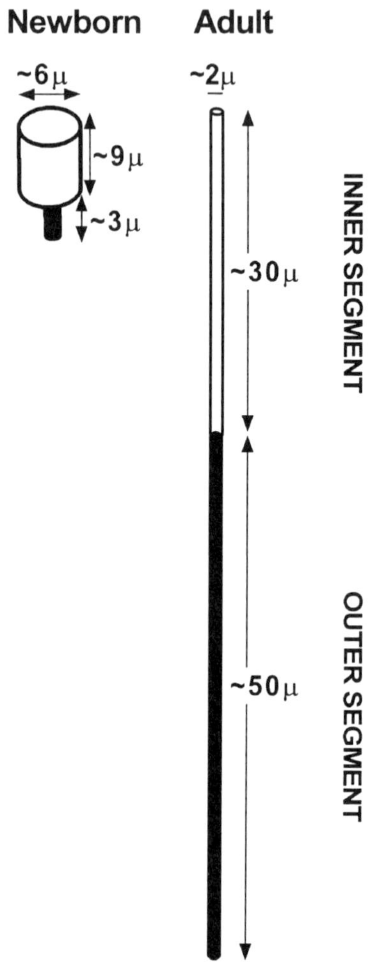

FIGURE 74.
Schematic rendering of a human cone illustrating the dramatic changes it undergoes between birth and adulthood. (From Banks & Bennett, *Journal of the Optical Society of America,* 5:2059-2078, 1988, with permission from the Optical Society of America).

Development of the dorsal lateral geniculate nucleus

All dLGN cells are born in the first quarter of gestation in both the human (Decaban, 1954) and primate fetuses (Rakic, 1977). At approximately the same time the dLGN is anatomically differentiated. Lamination begins at midgestation in both species and is quite complete at birth (Hitchcock & Hickey, 1980; Rakic, 1977). Optic tract fibers arrive and start to form synapses in primates dLGN by embryonic week 8 (Hendrickson & Rakic, 1977). Initially, the inputs from the two eyes overlap extensively. The segregation of inputs from the two eyes occurs prenatally in the rhesus monkey and certainly precedes the formation of ocular dominance columns in layer IV of the primary visual cortex (Rakic, 1976). A crude retinotopic map is already in place by the time that all the optic tract fibers have reached the dLGN. In humans, cell size in the parvocellular laminae reaches adult size within the first year (Hickey, 1977). Conversely, cell body size in the magnocellular layers continues to increase until shortly after the end of the second year (Garey & de Courten, 1983).

The shape of the dendritic tree of dLGN cells is fairly mature at birth. The number of dendritic spines, however, reaches a peak at birth in primates (Garey & Saini, 1981) and around the age of four months in humans (de Courten & Garey, 1982). Subsequently, spine numbers decline to reach adult levels around 9 months (in humans) (de Courten & Garey, 1982). The number of synapses follows a similar course, which is marked by a 50% reduction between birth and 8 weeks in the parvocelular layers and between birth and 16 weeks in the magnocellular layers in primates (Gottlieb et al., 1985). Functional cell classes in the primate retina and dLGN are clearly discernible at birth: almost all cells can be classified as either X or Y based on their spatial summation properties. The distribution of the two classes of cells across laminae is also adult-like.

Neural events involved in the development of retinotopic maps in the midbrain

A striking phenomenon in the development of the central nervous system is that the topographical arrangement of receptor cells is maintained in their projections to every station of the afferent pathway in question. This is true for the organization of afferent projections to the dLGN and superior colliculus in mammals and to homologous structures (optic tectum) in fish and amphibians. The optic tectum in these species plays a crucial role in visual localization, similar in many respects to the role of the superior colliculus in mammals. Researchers study the development of retinotectal projections in the hope of finding clues for the processes involved in the formation of retinotopic maps in mammals. Given that these projections are in place before the onset of visual experience, the emergence of topographical specificity in the developing visual pathway is likely to depend upon one of two processes (or both): 1) guidance of axons via chemical markers, and 2) competitive interactions among neurons based on spontaneous neuronal signaling. The formation of retinotopic maps in the central nervous system has been studied extensively in fish and frogs.

The orderly projections from the retina to the tectum ensure that the spatial layout of the images that fall on the retina is conveyed accurately to these midbrain structures. The integrity of this map is a prerequisite for the ability to localize and capture prey and to avoid predators (Springer et al., 1977). Retinotectal projections in fish and frogs remain highly plastic throughout the adult life in contrast to the corresponding projections in mammals. This property has enabled researchers to introduce drastic changes in the afferent pathways and observe how the system becomes reorganized. These manipulations involve the resection of parts of the retina or the tectum and the replacement of the resected parts with implants taken from non-corresponding parts of the donor's retina or tectum, respectively.

Normally, retinotectal projections develop in three major stages. In the *first stage*, first-order afferents grow along the optic nerve, enter the central nervous system, and arrive in the tectum in an orderly fashion. During this process, the growing retinotectal axons must overcome two major pathfinding problems. First, all afferents must cross the midline and continue their trip toward the contralateral tectum. Second, the afferents from the nasal hemiretina must find their way to the caudal part of the tectum, whereas fibers from the temporal hemiretina must terminate in the rostral part of the tectum. The *second stage* in the development of retinotectal projections involves the formation of synaptic contacts in the appropriate part of the tectum. The *third* and final *stage* involves the stabilization of the synapses formed with appropriate target cells in a way that establishes the point-to-point precision of the retinotopic map.

Processes involved in the first two stages of retinotectal development will be discussed first. Initially, it was thought that axonal guidance, during the *first stage* in the development of retinotectal projections, relies on the maintenance, by migrating axons, of the spatial arrangement of their cell bodies in the retina. A major problem with this proposal is that the relative spatial position of retinotectal fibers changes as they cross the optic chiasm. Researchers then started to examine the possibility that chemical interactions play a role in axonal guidance. Using monoclonal antibodies to target specific molecules, they identified a number of chemical agents, mostly surface proteins and lipids, which are distributed along the visual pathway in a topographically specific manner. Some examples are the topographically distributed molecule (TOP), the 9-0-acetylated ganglioside, and the neural adhesion molecule (N-CAM). The role of adhesion molecules in axonal guidance is consistent with the finding that the amount of such agents, measured in tissue samples taken at various levels of the visual pathway, correlates with the residual ability for topographical reorganization, following removal of portions of the retina or tectum. Chemoaffinity appears to operate at various levels of the visual pathway, including the retina, tectum, and among the growing axons. Chemical interactions seem to operate in different ways. *Selective adhesion* may allow axons that originate in adjacent retinal loci to recognize one another and grow in close spatial proximity. In addition, adhesion molecules located along the prespecified path of growing axons (as for instance, within the primitive hypothalamus and on glial cells, Mason & Sretavan, 1997), operate as *positional markers* and constrain the course of these axons. Finally, *interactions between chemical agents* located on the growth cones and complementary molecules in the tectum may establish gradients of increased or reduced preference for rostral or caudal tectal regions depending upon the axon's site of origin in the retina, during the *second stage* of retinotectal development. Each of the three processes will be examined separately.

The contribution of chemical affinity among growing axons to axonal guidance was examined in experiments in which the caudal portion of the tectum is removed and the rostral portion is resected, rotated by 180 degrees and put in place of the caudal part (Sharma, 1975). In this case, the regenerated axons form normal projections so that nasal afferents innervate tissue taken from the rostral tectum. It appears that fibers from adjacent regions in the nasal hemiretina were redirected in such a way that axons, which would normally project to the posterior part of the caudal tectum, ended up in the anterior part of the implanted caudal tectum, and vice versa.

The second guidance mechanism involves contact between axons and various structures located along their path (i.e., not necessarily other axons or target cells). At least in certain species, structures that are formed prior to the onset of axonal growth form boundaries that constrain the path of outgrowing axons and prevent them from following inappropriate alternative routes (Wu et al., 1988). One of these structures is the *glial knot*, a neuroepithelial region that forms the rostral boundary of the optic nerve at the level of the optic chiasm in the chick and mouse embryo.

The existence of the third guidance mechanism, one involving positional markers inside the tectum, is supported by in vitro studies. In these experiments retinal ganglion cell axons are allowed to grow on cultured membranes formed by alternating strips of rostral and caudal tectal cells. When no restrictions are imposed upon the growth of these fibers, axons from the temporal hemiretina consistently avoid caudal strips. In vivo studies complement these findings. For instance, when the caudal part of the tectum is resected and the optic nerve crushed, initially only axons from the appropriate part of the retina regenerate and project to the intact rostral tectum (Schmidt, 1983). Electrophysiological recordings conducted a few months later reveal, however, that axonal arbors from the other half of the retina move rostrally and form synapses with tectal neurons in the rostral tectum. Apparently, afferents that once terminate in the caudal tectum can be rerouted toward the rostral tectum. The spatial layout of these "inappropriate" projections shows orderly organization as if the entire retinotopic map was compressed to fit inside the intact half of the tectum. A similar conclusion can be drawn from studies in which the tectum is left intact and, instead, the nasal hemiretina is removed. Electrophysiological recordings have shown that axons from the intact retina slowly occupy the denervated rostral tectum (Schmidt et al., 1978). These findings indicate that positional markers in the tectum are, to a certain extent, responsible for the formation of the crude retinotopic map that is found during the early stages of tectal development. It appears, however, that rather than operating as a strong inhibitor for the growth of temporal axons toward the caudal tectum or, alternatively, as a strong attractor for nasal fibers toward the caudal tectum, chemoaffinity agents create gradients of *relative preference* along the tectal surface.

The first two stages in the formation of a retinotopic map in the tectum (i.e., migration of axons into the tectal area and synaptogenesis) do not appear to depend upon the presence of electrical activity in the retinotectal afferents. Thus, blockade of retinotectal activity with the Na^+-channel blocker tetrodotoxin does not prevent the initial emergence of an as yet crude retinotopic map in the tectum (Harris, 1984). The role of activity-dependent processes becomes crucial during the third and final stage, which involves the stabilization of retinotectal synapses and the refinement of the retinotopic map in the tectum. This stage is especially important, considering that the majority of the initial synapses formed by in growing axons are transient. Early in development in the tectum of fish and amphibians and in the dLGN in mammals, there is an abundance of aberrant collateral and branches, that are often longer than in the adult animal and cross into inappropriate laminae (LaChica & Casagrande, 1988). In addition, terminal branches of adjacent retinal ganglion cells in the tectum are significantly larger early on and show considerable spatial overlap (Schmidt & Eisele, 1985). Retinotopic map refinement, which requires elimination of exuberant branches, depends upon spontaneous activity of retinal ganglion cells. This is also true in mammals, where map refinement is well under way before the onset of visual experience, clearly implicating spontaneous activity in this process.

The role afferent activity plays in the refinement of retinotectal maps presumably depends upon the capacity of axons that originate in neighboring retinal loci to drive their tectal targets in synchrony. As shown in schematic form in Figure 75, initially there is extensive overlap between the terminal fields of retinal ganglion cells. A given tectal cell (like cell "F" in the figure) receives input from retinotectal fibers that originate in adjacent retinal loci (cells labeled "C" and "D") and also from fibers that innervate more remote regions in the retina (like cell "B"). Subsequent selective stabilization of certain synapses and elimination of others is responsible for the mature retinotopic maps. This process follows two general rules: 1) contacts which are effective in driving their postsynaptic targets are more likely to be retained and, 2) contacts which consistently fail to drive their target cells are likely to become eliminated. These conditions were originally formulated by Hebb (1949) to account for the processes of associative learning in mature organisms. What makes activity-dependent synaptic stabilization work is that neighboring fibers have the tendency to fire in synchrony and, therefore, are more likely to drive their common postsynaptic targets. This phenomenon starts before the onset of visual experience and, in some species, even before birth (Meister et al., 1991). Another requirement for the effectiveness of Hebbian rules is that the distribution of synaptic contacts from distant and neighboring retinal ganglion cells is not completely uniform, even before the onset of neuronal signaling in the retinotectal pathway. Growing axons have been guided in such a way as to preserve, to a certain extent, the relative positions of their cell bodies in the retina. This means that the density of initial synaptic contacts formed by neighboring retinal ganglion cells onto a given tectal cell is typically higher than the density of synapses formed by the afferents of more distant ganglion cells. At this point, chemoaffinity-dependent axonal guidance mechanisms have fulfilled their role, and activity-dependent mechanisms come into play. Tectal cells that receive denser synaptic inputs from neighboring cells in a particular retinal locus are more likely to be driven by those cells firing in synchrony than by afferents that originate in more distant retinal loci and make fewer synaptic contacts.

As we have seen in Chapter 4, optic nerve fibers that project to the dLGN in mammals are anatomically segregated into distinct classes that parallel those found among retinal ganglion cells. In mature animals, X and Y retinal ganglion cells project to different targets in the dLGN. A similar segregation is noted for ON-center and OFF-center cells. External visual stimulation does not appear to play a critical role in the development of functional cell class-specific segregation, at least at the level of the dLGN, as suggested by the lack of serious consequences of monocular lid suture and dark-rearing from birth.

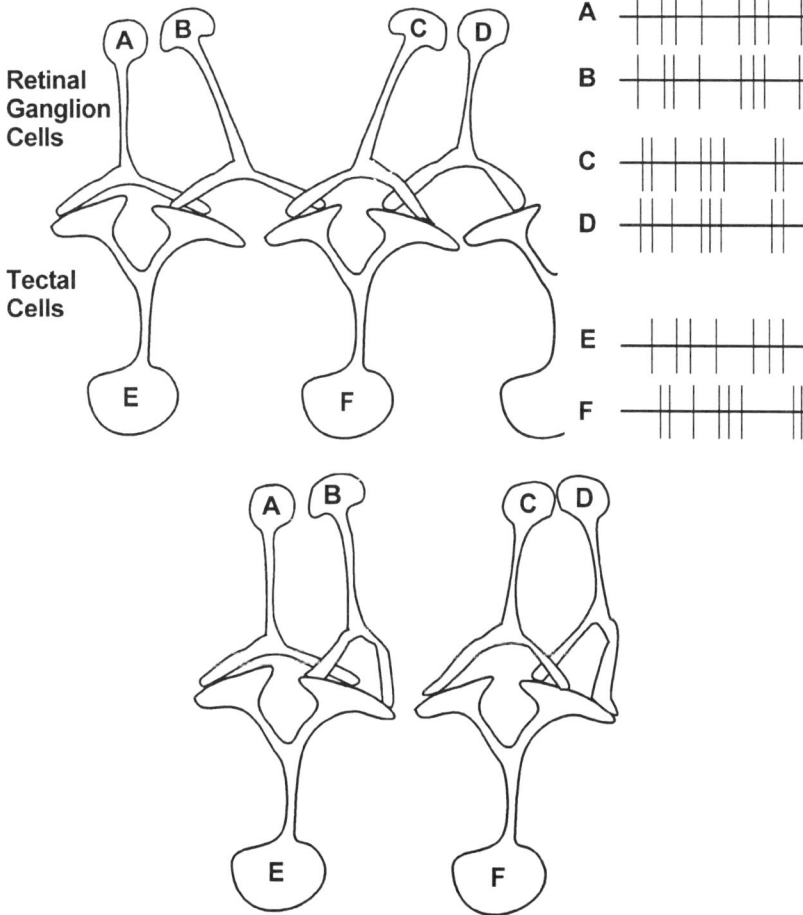

FIGURE 75.

Schematic rendering of retinotectal connections before the onset of spontaneous activity in retinal ganglion cells (top left) and after (bottom). Ganglion cells "C" and "D" are close neighbors in the retina, and their firing shows a high degree of intercorrelation (top right). This is also true for cells "A" and "B". In contrast distant cells "B" and "C" rarely fire in synchrony. Initially cells "C" and "D" form contacts with tectal cell "F" which also receives input from distant retinal cell "B". The combined synchronous input from cells "C" and "D" are sufficient to drive cell "F". As a result the correlation in the firing patterns between presynaptic cells "C", "D" and the postsynaptic cell "F" is also high. According to this model, this condition leads to the maintenance of the synapses made by cells "C" and "D" on tectal cell "F" and the elimination of "inappropriate" connections, like the one formed by cell "B" of cell "F" (and also by cell "C" on cell "E"). (From: "Activity sharpens the regenerating retinotectal projection in goldfish: Sensitive period for strobe illumination and lack of effects on synaptogenesis and on ganglion cell receptive field properties", Schmidt, J.T. & Eisele, L.E. *Journal of Neurobiology, 19,* 395-411. Reprinted by permission of Wiley-Liss, Inc., a subsidiary of John Wiley & Sons, Inc.).

In contrast, segregation of the inputs from ON-center and OFF-center ganglion cells can be prevented by complete blockade of activity in the optic nerve from birth to five weeks or later (via repeated tetrodotoxin injections into one eye; Dubin et al., 1986). The potential role of correlated firing was suggested by the finding that adjacent retinal ganglion cells having the same type of receptive field (i.e., either ON-center or OFF-center) have the tendency to fire synchronously. In contrast, cells that have different types of receptive fields are less likely to fire in synchrony, even when they sample the same portion of the visual field (Mastrogrande, 1983).

To summarize, the formation of the retinotopic map in the optic tectum in fish and frogs, and in dLGN in mammals, progresses in three main stages. These include: 1) migration of growing retinal ganglion axons into the central nervous system and arrival at the appropriate part of the target nucleus, 2) synaptogenesis, and 3) stabilization of synapses formed with the appropriate targets, a process leading to the refinement of the retinotopic map. The initial axonal guidance depends upon chemical agents that establish preference gradients among axons that originate from adjacent retinal loci, and also among afferents that originate in the nasal or temporal hemiretina and the caudal or rostral part of the tectum, respectively. A third possible mechanism may involve guidance of growth cones through direct contact with neighboring structures (not necessarily other axons). The retinotopic map undergoes substantial refinement postnatally. This refinement depends upon spontaneous neuronal activity and involves elimination of exuberant axonal branches and synapses. Correlated activity between the afferents that originate in adjacent retinal loci plays an important role in this process. A more detailed discussion of the molecular mechanisms underlying activity-dependent processes is deferred until the end of this chapter following a discussion of similar phenomena in the mammalian visual cortex.

Chapter 9

Maturation of the Visual Cortex

Anatomical development

The neocortical plate is formed by cells that migrate from a layer of precursor cells that line the surface of the ventricles in the fetus. During their migration, newly formed neurons have to cover an impressively large distance toward the developing cortical plate. Neuronal migration progresses in an orderly fashion from deep to superficial layers. Migrating cells find their way to their final destination by traveling along the elongated processes of radial glial cells (Rakic, 1972). Cell migration is performed in a spatially specific manner, so that throughout this process migrating cells retain their relative positions in the ventricular zone.

The ontogenesis of the striate cortex has been studied extensively in primates using autoradiographic tracers such as ^{3}H-Thymidine. When this compound is injected in pregnant monkeys, it travels through the blood stream and the placenta to the fetus where it selectively labels cells in the final stage of cell division. Injections are usually made at various times during gestation in different animals. Then the cells are allowed sufficient time to migrate to their final positions in the cortical plate. Brain slices from the newborn monkeys are then photographed to determine the position of labeled cells. Using this method, researchers have concluded that all cortical neurogenesis in primates is complete during the first half of pregnancy (reviewed by Rakic, 1991).

Most geniculocortical axons approach the cortical plate before all of the neurons that will later form layer IV have completed their migration. The incoming axons do not enter the immature cortical plate immediately, but form an intermediate zone underneath it, known as the *subplate*. Some researchers believe that the subplate serves as a "waiting" compartment for dLGN axonal terminals until cortical layer IV is formed (Shatz et al., 1990). Special subplate neurons project to the still forming cortical layers and appear to play a scaffolding role for incoming axons. The geniculocortical afferents start forming synapses with layer IV neurons around embryonic week 18 (Rakic, 1976).

After crossing the subplate, target selection by growing axons is guided by activity-dependent processes. Thus, intraventricular infusion of tetrodotoxin during the final stages of gestation in the cat leads to grossly aberrant projection and axonal arborization patterns within area 17 (Catalano & Shatz, 1998). It is, however, unclear whether neuronal signaling between incoming geniculostriate axons and subplate cells is directly involved in guiding target selection or if this activity simply facilitates chemoaffinity-based processes which in turn determine target selection. At birth the striate cortex already exhibits the adult pattern of lamination. Cortical patches or columns are also present prenatally. For instance, patches of cortex that become heavily stained by cytochrome oxidase in layers II and III have been found as early as embryonic week 21. It appears that although the number of cytochrome oxidase compartments remains constant during the course of postnatal development, the average size of each blob may increase by approximately 40% (Purves & LaMantia, 1993).

Perhaps the most impressive sign of postnatal growth in striate cortex concerns the number of dendritic spines and synapses. It is estimated that only about 10% of the total number of synapses that exist in the adult visual cortex were formed at birth. The number of dendritic spines (roughly corresponding to Type I or excitatory synapses) doubles in the period between birth and six months in humans (Mitchel & Garey, 1984) or eight weeks in primates (Booth et al., 1979). The total number of synapses (i.e., Type I and Type II synapses combined) shows a four-fold increase during the same period (Huttenlocher et al., 1982). In the magnocellular input layer (IVCa), the peak in spine density is noted somewhat earlier (at approximately five weeks; Lund & Holbach, 1991), and its timing correlates well with a reduction in the mean length of dendritic arbors in these layers.

Given that spiny stellate cells in lamina IVC are the primary recipients of geniculostriate inputs, the overproduction of spines may mark the peak of the critical period for synaptic reorganization which extends beyond the completion of eye-specific afferent segregation in this layer (LeVay et al., 1980). Moreover, it is reported that the time window for developmental plasticity phenomena related to ocular preference is longer in lamina IVCb compared to lamina IVCa (LeVay et al., 1980). This finding is consistent with the fact that the peak in spine density is reached later in the parvocellular (IVCb) than in the magnocellular input lamina (IVCa). Since the total number of cortical neurons does not change dramatically after birth, the reported increase in the number of synapses must largely be due to an increase in the number of connections per neuron. Visual experience is apparently one of the factors that affects the growth of neuronal processes and synapses in the visual cortex. For instance, an increase in the number of dendrites per neuron in area the visual cortex during development is found in rats reared in enriched environments (Greenough et al., 1985).

Regressive phenomena are also noted in both humans and primates. Thus, there is a modest reduction in the number of neurons between birth and adulthood, which is estimated to be in the order of 16% in primates (O'Kursky & Colonnier, 1982). At birth, neuronal density is already past its peak and has started to decline (Huttenlocher, 1990). The number of dendritic spines declines to reach adult levels at eight weeks. The decrease in the number of synapses per neuron and synaptic density starts after approximately eight weeks in primates and eight months in the human infant, reaching adult levels no earlier than 36 months in primates (O'Kursky & Colonnier, 1982). The corresponding age for humans is probably between two and three years.

Development of interlaminar connections

As already mentioned in Chapter 5, connections between cortical laminae (i.e., vertical connections) play an important role in the emergence of the complex response properties of cortical cells (Hubel and Wiesel, 1963). Receptive field layout and orientation selectivity of simple cells, especially in primates, is produced in part by converging input from cells with similar orientation preferences located in different cortical layers. There is also some evidence that the emergence of complex receptive field properties in striate cortex parallels the course of maturation of certain types of vertical connections. For instance, during the first two postnatal weeks only simple cells can be found in area V1. During the third week the number of connections formed by stellate cells in layer IV with cells in layers II and II reaches adult levels. At the same time, cells with complex receptive field properties are first encountered in layers II and III. It is thus possible that converging input from layer IV cells is a prerequisite for the development of complex receptive field properties by cells in supragranular layers.

There is accumulating evidence that the development of interlaminar connections in the mammalian primary visual cortex is characterized by specific growth of axonal branches in the appropriate target layers at, or very soon after, the axons reach their final target cells. For instance, Katz (1991) used the Lucifer Yellow intracellular staining technique to examine the branching patterns of individual pyramidal cells in newborn kittens. According to this report, collaterals from pyramidal neurons in layers II and III terminate almost exclusively within layers II, III (intralaminar connections), and layer V (interlaminar connections), as soon as these cells acquire efferent processes. As in adult animals, layer IV does not contain axon terminals from cells located in superficial layers. No evidence of elimination of primary collaterals was found at later stages of development. The axons of spiny stellate cells in layer IV also form specific connections with the appropriate target neurons in layers III and V before birth (Katz & Callaway, 1990; Lund et al., 1977). Katz and Callaway (1990, 1992) reported that continuous binocular deprivation during the first few postnatal months does not affect the capacity of spiny stellate cells in layer IV and of pyramidal cells in layers II and III to form specific interlaminar connections. Therefore, the processes through which appropriate vertical connections are formed in primary visual cortex do not appear to depend upon patterned visual stimulation. These findings are consistent with observations that the initial emergence of orientation preference in visual cortex does not depend upon visual stimulation (Wiesel & Hubel 1974).

One process that may be responsible for the specificity in interlaminar connections early in development may involve inhibitory chemical interactions among growing axonal processes and cells (neurons or glia) located in the target laminae. Chemical agents may, for instance, inhibit the formation of synapses between pyramidal cells in layers II and III and neurons in layer IV (Lund, 1988). This process would ensure that the afferents of cells in the supragranular layers are normally restricted within layers II, III, and V.

Development of intralaminar connections

Intralaminar (horizontal) connections are especially prominent in layers II, III, and V in both primate and in nonprimate species. In addition to short-range, local connections confined within the boundaries of a single cortical column, neurons form long-range connections with cells in different columns. These connections are formed by clusters of synaptic terminals located at regular intervals of approximately 1 mm along extensive axonal collaterals (up to 8 mm in length). Clustered synaptic contacts are usually formed with cells that show similar orientation preferences (e.g., Gilbert & Wiesel, 1989). The receptive fields of these neurons sample distinct but adjacent portions of the visual field. As already mentioned in Chapter 5, these connections are suitable for coding the orientation of linear contours in relation to surrounding visual patterns, a process which may be involved in perceptual context effects (Van Essen et al., 1991). In the mature visual system, clustered horizontal connections are probably responsible for patterns of synchronous firing among cells with similar orientation preferences that receive input from different visual field regions (e.g., Hata et al., 1991).

In the cat, crude synaptic clusters first appear in the supragranular layers in the immediate postnatal period, that is, before eye opening (Galuske & Singer, 1996). Neither patterned visual experience nor retinal activity is necessary for the initial emergence of crude clusters, which is not prevented by binocular deprivation, dark rearing, or intraocular injection of tetrodotoxin (reviewed by Katz & Callaway, 1992). Initial clustering of synapses can, however, be severely disrupted by blocking cortical activity, suggesting that neuronal signaling among neurons plays a key role in this process (Ruthazer & Stryker, 1996). In primates, spontaneous neuronal activity must play a somewhat greater role in the formation of crude synaptic clusters, given that, at the onset of visual experience, such clustering is at a far more advanced stage than in the cat.

Subsequent maturation of clustered connections involves a dramatic increase in the total length of collaterals (possibly in order to adapt to the expansion of cortical surface with age) and cluster refinement which, in the cat, is essentially complete at the age of 4-6 weeks (Luhmann et al., 1991). The latter process involves, in turn, selective elimination of terminal branches located between iso-orientation patches (Callaway & Katz, 1990), and possibly the formation of new clusters, as shown in Figure 76. The addition of new clusters is necessary in order to maintain the largely constant inter-cluster spacing observed throughout development. The final stage in the shaping of long-range clustered connections is controlled by processes that depend upon visual-experience, given that binocular deprivation starting at the end of the third week essentially "freezes" further elaboration of these connections (for a review, see Schmidt et al., 1999). More on the development of intralaminar connections in relation to other aspects of functional neuronal properties follows in the next section.

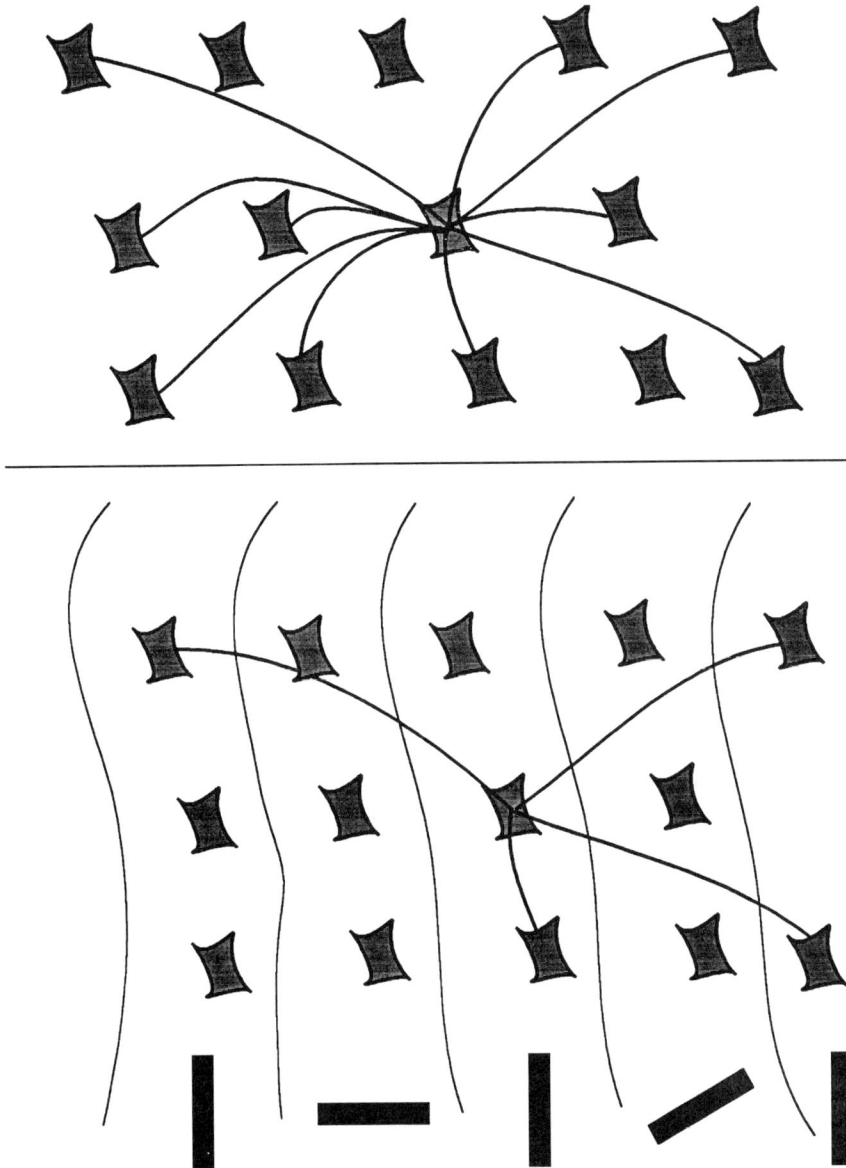

FIGURE 76.

Schematic diagram of intralaminar connections in layers II/III. At birth (top) only crude clusters are formed: cells like the one shown in the middle of the array project to nearby neurons in a semi-random pattern. During the first few weeks after eye-opening, synaptic clusters become well-defined and restricted within iso-orientation columns. For simplicity, only a few orientations are shown in this representation of the cortical surface. (From Katz & Callaway 1992. With permission, from the *Annual Review of Neuroscience, Volume 15.* Copyright 1992 by Annual Reviews www.AnnualReviews.org).

Development of spatial resolution and orientation selectivity in visual cortical neurons

The improvement of spatial resolution, displayed by individual cells in the foveal representation of the striate cortex follows, a parallel course with the maturation of the spatial resolution of dLGN cells (Blakemore & Vital-Durand, 1986b). Moreover, the rate of improvement in spatial resolution of the "best" cells in the dLGN and primary visual cortex, closely matches the developmental course of behaviorally measured acuity. However, early in development behavioral acuity estimates are consistently lower than the best spatial resolution displayed by individual neurons. This finding probably reflects the contribution of nonsensory factors (more details are given below).

During the first postnatal weeks, striate cells in monkeys reared with both eyelids sutured closed at birth show a considerable degree of functional maturation. The great majority of cells show preference for a restricted range of stimulus orientations. In addition, a regular progression of orientation preferences is encountered along oblique electrode penetrations (Wiesel & Hubel, 1974). Overall, the organization of cells according to their orientation preference approaches that of the adult animal. Identical results were obtained from a monkey tested only two days after birth. These findings suggest that extensive visual experience is not crucial for triggering the processes that lead to the emergence of orientation selectivity in primates.

In the cat, orientation preference develops, initially, independent of visual experience. Neurons that show preference for a restricted range of visual contour orientations can be found immediately after eye-opening (Hubel & Wiesel, 1963), and a regular pattern of iso-orientation patches can be seen using optical imaging by day 14 (Crair et al., 1998). As shown in Figure 77, binocular pattern vision deprivation does not severely impact the development of cortical orientation-preference maps during the first 2-3 postnatal weeks (Gödecke et al., 1996). After this period, and under conditions of normal visual experience, there is a significant improvement in both the proportion of cells that show narrow orientation preferences and in the degree of selectivity of individual cells (for a review, see Fregnac & Imbert, 1984). Binocular deprivation that starts at this age prevents further maturation of orientation selectivity and leads to a further deterioration of cortical maps for orientation preference (Crair et al., 1998). These findings suggest that patterned visual experience is necessary, at least for the maintenance of orientation sensitivity.

Moreover, there is ample evidence that the specific content of visual experience not only is critical for the maintenance of orientation bias that emerges prenatally, but also contributes to the sharpening of orientation tuning. For instance, it has been shown that the distribution of orientation preferences among visually responsive cells in the striate cortex of the kitten can be significantly biased by specific visual experience (Hirsh & Spinelli, 1970; Stryker et al., 1978). In these studies, kittens were reared wearing goggles, starting after the fourth postnatal week, so that each eye viewed large stripes of a single orientation, either horizontal or vertical. The distribution of orientation preferences, among the cortical cells studied, had shifted toward the orientation of the stripes viewed by the cells' dominant eye. For instance, cells that were dominated by the left eye were significantly more likely to show preference for vertically oriented stimuli in those animals that had been exposed to vertical stripes through that eye.

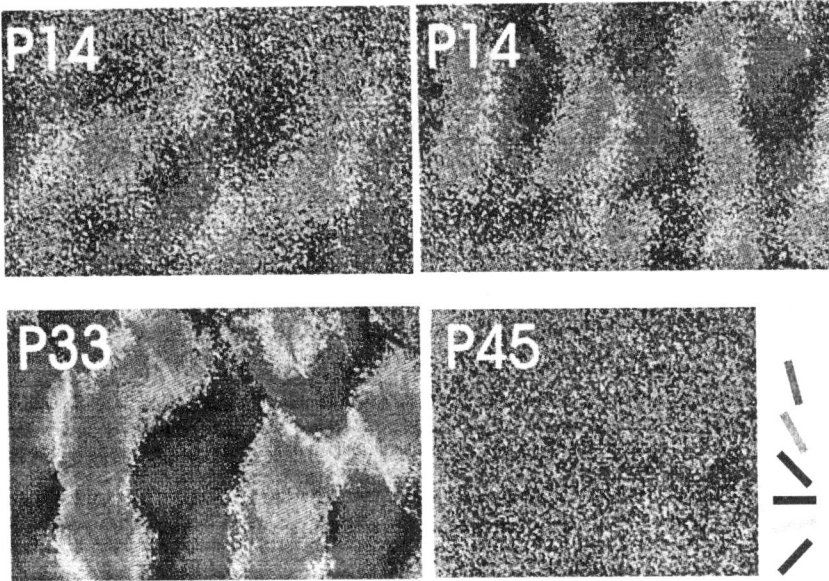

FIGURE 77.
The spatial layout of iso-orientation maps obtained using optical imaging in normally reared cats at two different postnatal ages (*upper left*: 14 days; *lower left*: 33 days). Notice the presence of clearly defined iso-orientation patches shortly after eye opening that become fully mature 2 weeks later. Corresponding maps appear very similar in binocularly deprived kittens early on (*upper right*). With prolonged deprivation, maps deteriorate significantly (*lower right*). Only maps produced in response to stimulation of the eye contralateral to the exposed cortex are shown here. (Reprinted with permission from Crair et al., The role of visual experience in the development of columns in cat visual cortex. *Science, 279*, 566-570. Copyright 1998 by The American Association for the Advancement of Science).

However, the overall number of cells whose orientation preference matched the orientation of the visual stimulus did not change, and a significant proportion of neurons that formed patches along the cortical surface did not respond to any orientation. These findings are consistent with the view that, in the absence of linear contours of a particular orientation, cells with an inborn predisposition toward that orientation are not consistently driven and eventually lose their sensitivity. This notion, which became known as the *selective hypothesis*, can explain why the number of orientation selective cells (in cats that were simultaneously exposed to two different stripe orientations) was much larger when compared to animals exposed to the same orientation through both eyes (Stryker et al., 1978). However, more recent findings suggest that not only can the orientation preference of individual neurons be modifi-ed (Fregnac et al., 1992) but also that extensive reorganization in the layout of iso-orientation maps can and does occur following exposure to a restricted visual environment (Sengpiel et al., 1999). These findings suggest that visual experience may also exert *instructive* influences on the functional organization of the visual cortex.

Finally, it is noteworthy that the effects of selective visual experience depend, to a great extent, upon interocular interactions. Thus, when infant primates were reared with one eye exposed to lines of a single orientation, while the other eye was kept closed by lid suture, the cells that were driven by the non-deprived eye showed a normal distribution of orientation preferences (Carlson et al., 1986). The cells that were driven by the deprived eye did not, as a rule, respond to the stimulus orientations to which the other eye had been exposed. These findings suggest that restricted visual experience can play a role in the development of orientation selectivity only if competitive interactions between the inputs from the two eyes are maintained.

At the cellular level there is, at present, little information regarding the nature of the processes that mediate neuronal activity- and experience-dependent developmental changes. There are indications that both the *initial* emergence of clustered horizontal connections and orientation selectivity, follow parallel time courses and are affected by common processes. For instance, neither process is affected by blockade of retinal activity (Ruthazer & Stryker, 1996). During this early stage, processes that could play a role in development include the following: 1) chemical interactions between geniculocortical afferents and their target cells in layer IV, and also between neurons in the non-afferent layers and their target cells within the same layers, 2) intrinsic activity patterns involving visual cortex neurons; this possibility is suggested by reports that blockade of such activity prevents the emergence of crude synaptic clusters (Ruthazer & Stryker, 1996), and 3) patterns of geniculocortical input which are, in turn, determined by retinal activity and shaped through feedback connections from the visual cortex (Weliky & Katz, 1999). These alternatives are not mutually exclusive. For instance, chemical interactions may be facilitated by neuronal activity during the early stage of cortical map formation.

During the next stage of development (which begins one week or so after eye opening in the cat) retinal input assumes a dominant role in the refinement of neuronal properties and cortical maps. Although, at this stage, the role of retinal activity is primarily to convey visual input, it can be manipulated independently of visual experience. Studies designed to examine the role of neuronal activity have shown that both intrinsic cortical activity (Chapman & Stryker, 1993) and specific patterns of retinal activity are crucial. Thus, when all inputs from one eye to the visual cortex are synchronized on a daily basis, even for brief periods of time (by electrical stimulation of the optic nerve), orientation selectivity is prevented from reaching mature levels, although iso-orientation maps are not abolished (Weliky & Katz, 1997). Clearly, intraocular activity patterns are important for the maturation of orientation tuning. More recent studies have examined specific patterns of activity more closely. For instance, disrupting the balance of inputs originating in ON- and OFF-center retinal ganglion cells, through selective discharge suppression of ON-Center cells, prevents further development of orientation tuning properties in cortical neurons (Chapman & Gödecke, 2000).[1] It is noteworthy that the general layout of iso-orientation maps did not change as a result of these manipulations (see Figure 78). These results are consistent with the notion that activity-dependent processes, which may require competition between geniculocortical afferents, play an instructive role during the final stage of maturation. Normally, segregation of receptive field subregions that primarily receive input from ON- or OFF-center afferents, develop progressively on a substrate of crudely clustered long-range intrinsic connections, under the guidance of an activity-dependent process.

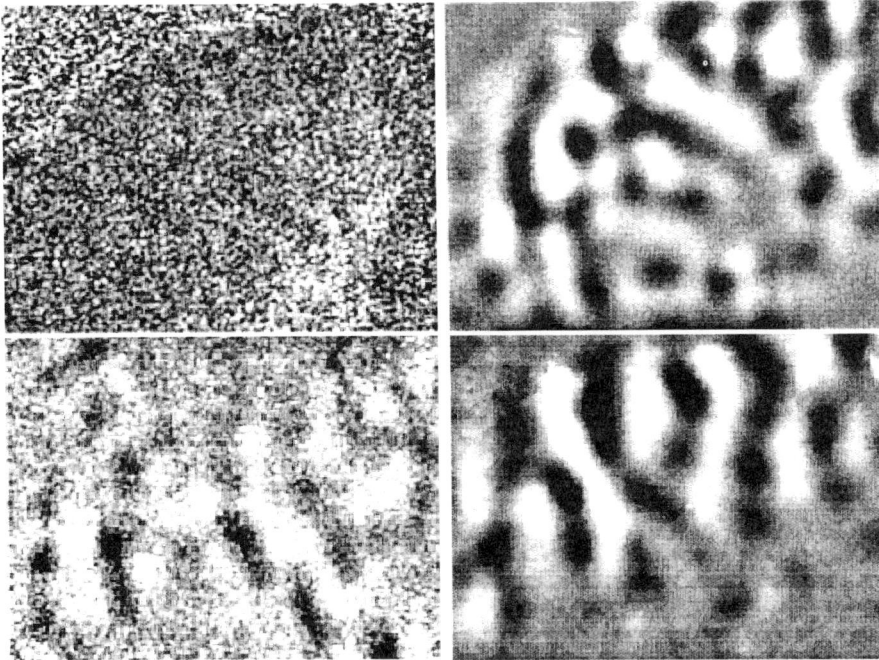

FIGURE 78.
The effects of blockade of ON-center cell input imposed at three different postnatal ages on the layout of iso-orientation maps in the primary visual cortex of the ferret. The maps shown were elicited by stimulation with a horizontal grating. Blockade prevents the emergence of the map when imposed soon after eye opening (postnatal day 21; upper left). Refinement of the map is arrested by blockade imposed during later stages of development (i.e., on postnatal day 28; lower left). No apparent deterioration of the cortical map occurs by blocking ON-center activity when the maturation of the map is nearly complete (i.e., on postnatal day 42; upper right). A normal map obtained at postnatal day 51 is also shown for comparison (lower right). (From Chapman & Gödecke, 2000, *The Journal of Neuroscience*, 20(5). Copyright 2000 by The Society for Neuroscience).

In conclusion, normal visual experience plays a crucial role in both the maintenance of already emerging orientation preferences and in the facilitation of further development of orientation selectivity in cortical neurons. Interactions between innate, experiential, and maturational factors define the conditions that permit the refinement of receptive field properties of visual cortex neurons. The time course and duration of the period of susceptibility to external input, may also be influenced by experiential and maturational factors (Aslin, 1981). Longitudinal data are particularly relevant to these issues, allowing researchers to examine changes in the response properties of individual neurons in animals reared under normal or experimentally altered visual experience. Up to this point, several key aspects of the structural and functional development of the visual system have been reviewed. However, as stated in the Introduction, the goal of sensory physiology is to establish relations between neural and perceptual phenomena. In the next section we examine how the maturation of the visual system affects the development of the ability to resolve fine detail during infancy. Specifically, our discussion focuses on the constraints imposed by the state of development of the visual system on spatial resolution.

The contribution of peripheral and postreceptoral factors in the development of spatial resolution in infancy

Measures of visual acuity and contrast sensitivity reflect the ability of the visual system, as a whole, to resolve fine details in the visual array. In the human newborn infant, visual acuity estimates obtained with behavioral methods range between 0.5 and 1 cycles/degree and reach 6 cycles/degree by the age of six months (Courage & Adams, 1990). Using visual evoked potential methods, which can presumably provide more direct measures of spatial resolution at the neuronal level, the estimates are substantially higher (Sokol, 1978). Both procedures reveal a rapid growth of spatial resolution in the first six months, continuing at a slower rate for several years. When adult levels of acuity are finally reached, they represent a near 40-fold increase over the estimates of visual acuity at birth. A similar developmental course is found in primates. In the following paragraphs we focus on the neural changes associated with the improvement of spatial sensitivity during early infancy.

In animal experiments, the spatial resolution of individual cells can be tested by presenting black and white gratings, that vary in spatial frequency and contrast, within the cell's receptive field and record the cell's discharge. When the spacing between the black bars is very narrow, the center and the surround of the receptive field receive equal amounts of light, as illustrated in Figure 79. Thus, the excitatory input triggered by stimulation of the receptive field center (in an ON-center cell), is canceled out by the inhibitory effect of the surround. As the distance between bars approaches the diameter of the receptive field center, those ON-center cells, whose center is covered by a bright strip (while most of their surround area is covered by the two flanking dark bars), will show an increase in discharge rate.

FIGURE 79.

Schematic representation of the change in receptive field size of a typical ON-center Pb cell between birth (*left*) and maturity (*right*). Only when the diameter of the receptive field center approximates the width of the white bars can the cell respond reliably to the grating stimulus.

For a given cell, one can determine a spatial resolution threshold that corresponds to the highest spatial frequency at which the cell's discharge first shows a significant increase above the rate of spontaneous firing. In adult primates, the best spatial resolution thresholds among dLGN neurons (typically Pß cells) closely matches the psychophysical acuity thresholds in this species (Crook et al., 1988). As previously mentioned in Chapter 4, the diameter of the dendritic tree of retinal ganglion cells sets an absolute limit to the spatial resolution of visual neurons. In a sense, the size of the dendritic tree corresponds to the sampling aperture of the ganglion cell. In the primate central fovea, there is a one to one correspondence between cone receptors and Pß retinal ganglion cells. Therefore, the spatial resolution of these cells is constrained almost exclusively by cone spacing.

During the early stages of postnatal development in primates, one finds a progressive reduction in the size of the receptive field that occurs primarily in dLGN cells that receive input from the fovea. The spatial resolution of these cells shows a dramatic improvement (from about 5 cycles/degree to 35 cycles/degree in the "best" cells) during the first 30 weeks of life (Blakemore & Vital-Durand, 1979). Similarly, the maximum spatial resolution of individual cells in the primate visual cortex increases by a factor of seven during the first postnatal year (Blakemore & Vital-Durand, 1983). In contrast, the spatial resolution of cells that receive input from the periphery of the retina changes little from birth to adulthood.

The factors that may contribute to the improvement of spatial resolution can be classified into two general categories: *peripheral* and *postreceptoral* (see Table 4). The former includes growth of the eyeball (indicated by axial length), improvement of optic media quality, and an increase in foveal cone density. It appears that purely optical factors cannot account for a substantial degree of improvement in spatial resolution. The net increase in axial length (only about 50%) cannot account for much of the change in spatial resolution (which increases by a factor of seven during the same period). Optical quality is also quite good soon after birth. On the other hand, retinal immaturity appears to be an important limiting factor for spatial resolution in infancy. The inner segment of foveal cones is a very inefficient waveguide for the incoming light because of its structural immaturity (Banks & Bennett, 1988). As a result, the funneling of light to the outer segment depends primarily upon its diameter. The latter is very slim at birth with a diameter that is approximately two thirds of the adult size (see Figure 74). An even more serious limiting factor is the average spacing of cones, which decreases by a factor of four between birth and adulthood. This change corresponds to a three-fold increase in the Nyquist limit (see Table 4). By taking into account recent estimates of the effective aperture and average cone spacing, Banks and Bennett (1988) surmised that the effective collecting area of cone receptors may change from 2% of the total retinal surface in the newborn to 65% of that area in the adult. Moreover, the ability of the outer segment of foveal cones to produce an isomerization in response to light may be as much as 10 times lower than in the adult.

Based on these estimates, Banks and Bennett (1988) computed the expected contrast sensitivity functions for the newborn "ideal"[2] observer and made comparisons with corresponding functions computed for the adult "ideal" observer.

Table 4. A comparison of optical and receptor characteristics in the
neonatal and mature eye.

Measure	Newborn infant	Adult
Axial length	16.6 mm	24.0 mm
Pupil diameter	2.2 mm	3.3 mm
Receptor spacing	1.66 min. arc	0.58 min. arc
Nyquist limit	20.9 cycles/degree	59.7 cycles/degree

Adapted from Banks & Shannon (1993).

They concluded that the amount of spatial detail that is made available to the central nervous system is approximately 19 times higher in the typical adult observer compared to the typical newborn infant. In conclusion, the constraints posed by peripheral factors can account for a significant portion of the discrepancy in visual acuity between the infant and adult observers (i.e., an approximately 30-fold difference). Peripheral factors cannot, however, account entirely for these differences. The discrepancy between the expected spatial resolution (given the estimated peripheral conditions) and the observed spatial resolution (indicated by visual acuity and contrast sensitivity measures) of newborn infants is due to postreceptoral factors.

Certain postreceptoral factors can be directly linked to functional and structural changes in the pathway that transmits neural signals from the retina to the visual neocortex. Other factors cannot be readily associated with an underlying change in neuronal organization. An example from the former category is the four-fold increase in the maximum discharge rate of individual units in the dLGN that is probably caused by a corresponding improvement in synaptic efficiency. This change alone could account for at least a two-fold improvement in spatial resolution (Blakemore & Vital-Durand, 1986). A second neural factor that may contribute to the developmental improvement in spatial resolution is a progressive reduction in the size of the receptive field center of dLGN cells. This, in turn, may be caused by a decrease in the number of receptor elements that converge upon individual retinal ganglion cells. Factors which cannot be directly associated with neural processes may include: inability to maintain attention toward the task-relevant stimuli and lack of motivation for producing a behavioral response, which is used as an index of stimulus detection in most infant psychophysical procedures. Moreover, very young infants may lack the readiness to produce a discriminatory motor response, such as a selective head turn toward a "target" stimulus, even when they can detect the experimental stimulus.

Segregation of eye-specific inputs to the visual cortex

In the adult cat and monkey virtually all neurons in layer IVC of the primary visual cortex respond to visual stimulation of a single eye (monocular cells), a phenomenon known as *ocular dominance*. The great majority of neurons in the remaining layers receive input from both eyes (binocular cells), but they tend to respond more vigorously when a visual stimulus is presented to one eye (i.e., they show relative preference for one eye). The initial segregation of geniculocortical afferents is a prerequisite for the establishment of binocular response profiles in non-afferent layers. It is believed that binocular response properties are the result of the convergence of two or more cells from layer IVC, which are driven by different eyes, onto individual neurons in the non-afferent layers. The remaining portion of this chapter is devoted to an examination of the conditions under which eye-specific afferent segregation develops. The focus shifts on these properties for two main reasons. First, this aspect of cortical organization is a prerequisite for the emergence of cortical binocularity which is, in turn, a necessary condition for the ability to perceive depth from stereoscopic cues (such as retinal disparity). Second, the establishment of eye-specific segregation is a valuable model for the study of the maturation of sensory cortices in general.

In primates, the segregation of geniculocortical fibers that carry input from different eyes begins in utero and follows the establishment of eye-specific layers in the dLGN (Rakic, 1976). During the first postnatal week, inputs from the two eyes in layer IVC still show extensive spatial overlap (see Figure 80). This has been demonstrated with histological (autoradiographic) and electrophysiological methods (LeVay et al., 1980). At the beginning of the second postnatal week, cells that are driven predominantly by only one eye begin to form clearly defined clusters. By the third week, injections of retrogradely transported fluorescent tracers (^{3}H-proline) into one eye produce a clear pattern of alternating labeled and unlabeled stripes visible in tangential sections through layer IVC. Even at this age, however, electrophysiological recordings reveal a small number of cells in layer IVC that are driven by both eyes. At six weeks, the pattern of alternating autoradiographically-labeled columns is similar to those found in adult animals. At that time, electrophysiological data revealed an equally sharp alternation of clusters of strictly monocular cells. In the cat segregation of geniculocortical afferents in layer IVC begins somewhat later. At birth, afferents are completely intermixed (LeVay et al., 1978). Physiological evidence of segregation (by combined optical imaging and electrophysiological recordings) can be found as early as the end of the second postnatal week, that is shortly after eye-opening (Crair et al., 1998). The adult pattern of ocular dominance patches is established by 8-10 weeks. Considerable functional and anatomical segregation can occur under conditions of form vision deprivation, including lid suture (Blakemore & Van Sluyters, 1975; Stryker & Harris, 1986) and dark rearing imposed at birth (Stryker & Harris, 1986). Thus, the initial stages of eye-specific segregation progress independently of patterned vision. Processes that operate before the onset of visual experience (i.e. in utero in primates and before eye-opening in the cat) must be responsible for the initial emergence of ocular dominance columns. It appears that these processes depend upon spontaneous retinocortical activity (Stryker & Harris, 1986), although chemical interactions among ingrowing geniculocortical axons and cortical neurons, involving chemoaffinity and selective adhesion, are also important, especially at the early

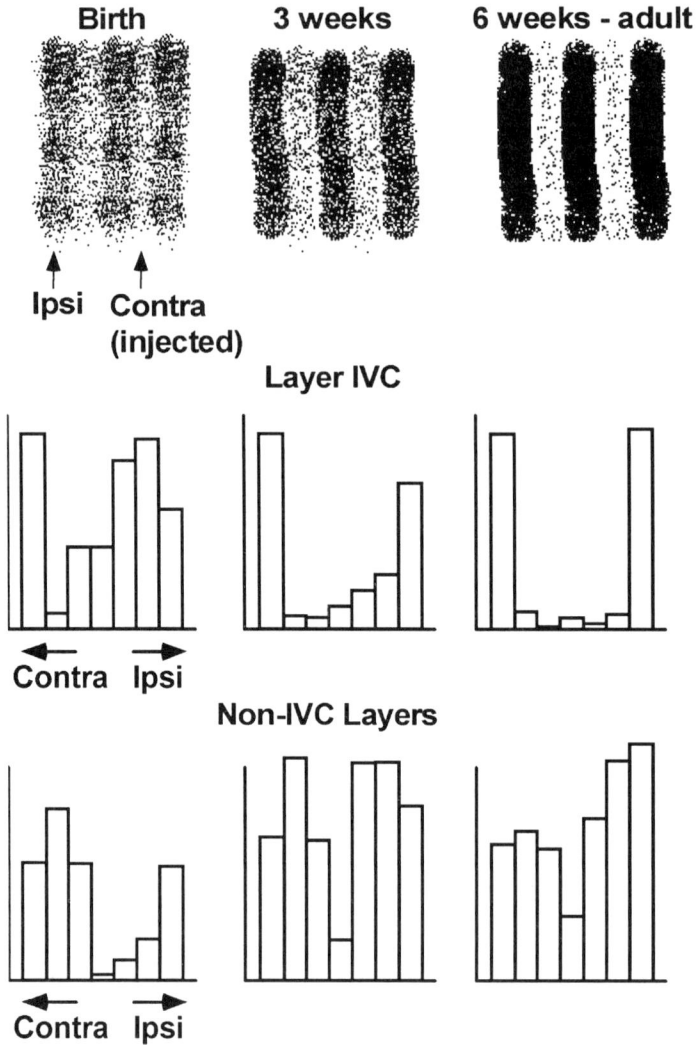

FIGURE 80.

Top: Schematic drawing of the staining pattern seen in tangential sections through lamina IVC following injection of ^{3}H-Proline into the contralateral eye at three different ages. Corresponding ocular dominance histograms, determined electrophysiologically for cells in layer IVC (middle) and in all other layers combined (bottom), are also shown. (Based on data from LeVay et al., 1980).

stages (Crowley & Katz, 1999). As we will see below, chemical interactions could be influenced by levels of neuronal signaling between geniculocortical axons and their target cells in lamina IVC.

Patterned vision is crucial for the refinement, stabilization and maintenance of the pattern of ocular dominance in layer IVC. A valuable tool in the study of the processes responsible for the later stages in the maturation of ocular dominance in layer IVC, and for its maintenance, involves the study of the effects of experimental manipulations of visual experience. The goal of these investigations is to determine how alterations of specific forms of visual experience can affect the course of segregation once it has been initiated. Data from deprivation experiments provide important clues regarding the role of functional interactions among geniculocortical afferents during cortical development. Early deprivation of one eye from patterned vision, by suturing the eyelids or by fitting an opaque eye patch (a treatment known as *monocular pattern-vision deprivation*), results in marked shrinkage of autoradiographically labeled patches in layer IVC. Labeling is accomplished by injecting ^{3}H-proline into one eye, the normal eye in Figure 81. Electrophysiological recordings performed after the deprived eye is opened, generally corroborate the autoradiographic findings. They reveal a marked expansion of regions containing cells that respond exclusively to stimulation of the non-deprived eye. Weaker, but still visible, deprivation effects are observed when monocular deprivation is imposed at an age at which the adult-like pattern of ocular dominance is nearly established (i.e., between 5 and 10 weeks; Horton & Hocking, 1997; LeVay et al., 1980). In the macaque, the sensitive period for monocular deprivation effects in layer IVC ends around the age of 10-12 weeks. Deprivation-induced shrinkage of ocular dominance patches is generally more severe in the parvocellular (IVCβ) than in the magnocellular (IVCα) input lamina (Horton & Hocking, 1997). Concurrent staining with ^{3}H-proline and cytochrome oxidase reveals severe shrinkage of blob regions that receive input from the deprived-eye in layers II/III. These changes parallel the deprivation-induced effects observed in layer IVC.

In the cat, susceptibility to monocular deprivation extends from the time of eye opening (i.e. around 9 days) to approximately four months (Hubel & Wiesel, 1970). The degree of susceptibility, however, varies widely during the sensitive period. Maximum susceptibility is found between the third and sixth postnatal week, a period during which monocular deprivation imposed for as little as two days can induce measurable physiological and anatomical changes (Olson & Freeman, 1975).[3] Recent data suggest that the shift in ocular dominance may start first in the non-afferent layers, even after very brief periods of monocular deprivation imposed at the peak of the sensitive period (Trachtenberg et al., 2000).

The degree of recovery, both at the cellular and behavioral level, following monocular deprivation depends upon the amount of interocular competition permitted by the experimental protocol (Sherman et al., 1974; Sherman & Guillery, 1976; Kratz et al., 1976). This can be achieved by manipulating the degree of relative advantage in favor of the previously deprived eye. The opening of the deprived eye can be paired with complete removal of the non-deprived eye. This treatment is known as *monocular enucleation* and induces a complete reversal of the relative advantage of the non-deprived eye. In other cases, the lids of the non-deprived eye are sutured (partial reversal of the advantage of the latter since diffuse light stimulation is still possible through that eye). Finally, opening of the deprived eye can be performed without interfering with the non-deprived eye (no extra advantage given to the deprived eye). The recovery of the ability of the deprived eye to drive cortical cells is maximal when the non-deprived eye is completely removed and minimal when the deprived eye is simply opened without imposing any deprivation treatment on the previously non-deprived eye (Kratz et al., 1976). When the competition exerted by the non-deprived eye is completely eliminated (i.e., by monocular

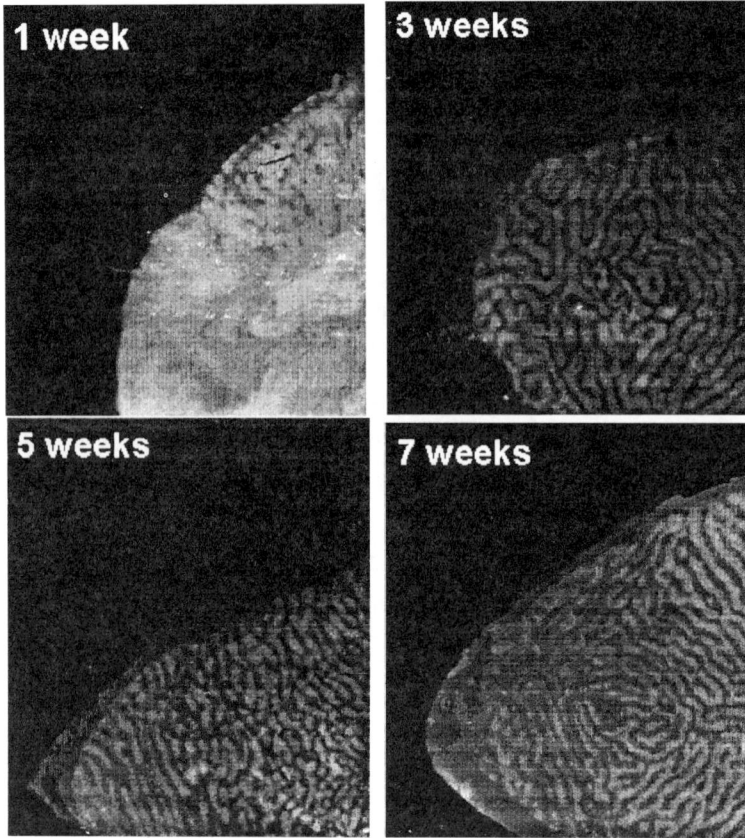

FIGURE 81.
Autoradiographic images of flat-mounted portions through layer IVCβ in four monocularly deprived monkeys. The images illustrate the distribution of ^{3}H-Proline (bright regions) that was injected into the normal eye and transported anterogradely into the visual cortex. Deprivation by lid suture was imposed at four different ages ranging from 1 to 7 weeks. The width of patches dominated by the non-deprived eye is maximal when lid-suture was imposed at 1 week and barely noticeable in the animal that was lid-sutured at 7 weeks. (From Horton & Hocking, 1997, *The Journal of Neuroscience, 17*. Copyright 1997 by The Society for Neuroscience).

enucleation) the afferents from the previously deprived eye acquire a clear advantage over the inputs from the previously non-deprived eye. If interocular competition is not reversed (i.e., when normal vision through the previously non-deprived eye is allowed) then the ability of the deprived eye to drive cortical cells shows minimal recovery. In that case, the capacity of the originally deprived eye to drive cells in lamina IVC can be reported as early as 12 hours following enucleation of the non-deprived eye (Kratz et al., 1976).

Monocular deprivation is accompanied by changes in the branching patterns of the geniculocortical afferent terminals in layer IVC. These changes can occur as early as six days following monocular lid suture in kittens (Antonini & Stryker, 1993a). Labeling geniculocortical efferents anterogradely, using the phaseolus lectin, revealed that the number of terminal axonal branches and the length of axons that carry input from the deprived eye, were significantly reduced, when compared to axons served by the non-deprived eye and to those found in normally reared kittens. In addition, the complexity of axonal arbors was significantly reduced in afferents served by the deprived eye. Thus, changes in the branching patterns of geniculocortical axons in layer IVC, following visual deprivation, occur at a much faster rate than originally thought possible, but it is not yet known if the timing of structural alterations closely parallels the course of physiological changes. These findings leave a number of unanswered questions. First, to what extent do structural changes depend upon functional interactions between the afferents from the two eyes? Second, do structural changes accompany the recovery of responsivity to stimulation of the deprived eye following reversed deprivation? For instance, it has not yet been shown if terminal branches of the afferents from the deprived eye can actually grow back following the reversal of the deprivation. However, assuming that regrowth does occur, it is uncertain whether it would be sufficient to fully restore the stimulus-specific properties and the precise alignment of the monocular receptive fields found in non-deprived animals (Antonini & Stryker, 1993a).

A likely scenario for the course of ocular dominance column formation in IVC (e.g., LeVay et al., 1978) involves, initially, a marked proliferation of synaptic terminals in this lamina formed by geniculocortical inputs immediately before the beginning of segregation. At this time, axonal arbors are longer than in the adult animal. Segregation proceeds by gradual selective elimination of synaptic terminals that leaves clusters of synapses in regions that will later become the centers of ocular dominance columns. Deprivation experiments strongly suggest that this process is locally mediated by two general processes: *competition* between the afferents that carry inputs from different eyes, *and cooperation* between the afferents that originate in the same eye. The outcome of these processes depends upon activity in presynaptic neurons and is ultimately determined by the degree to which each set of eye-specific afferents are successful in inducing postsynaptic activity in their target neurons in lamina IVC. It is also likely that competition by neighboring eye-specific terminals for specific molecules is crucial for their growth and/or their survival (Guillery, 1988). Competition would be especially intense during those periods of early development, which are characterized by an overabundance of neurons and neuronal processes and a limited supply of those agents. A variety of compounds, mostly neurotrophins, have been identified. They are released from cortical cells and may promote the maintenance and growth of synaptic connections. Competition could be influenced by activity dependent processes. A more detailed discussion of neurotrophic agents is reserved for a subsequent section, following the discussion of the role of neuronal activity in the formation of ocular dominance columns, which is presented below.

The role of activity-dependent cellular mechanisms in eye-specific segregation of geniculocortical afferents

As mentioned in previous sections of this chapter, activity-dependent mechanisms have been implicated in a number of phenomena in neural development, including the refinement of the retinotopic map in the tectum in fish and frogs (e.g., Schmidt, 1994), and the segregation of inputs from ON- and OFF-Center ganglion cells in the dLGN of the cat

(Dubin et al., 1986). In mammals, two processes have been the focus of extensive research: 1) the formation of ocular dominance columns in layer IVC and, 2) the emergence of binocular neurons in the non-afferent layers of the visual cortex. It is suggested that the segregation of geniculocortical afferents and the emergence of cortical binocularity follow the same set of rules, operating at different developmental stages. In this section the development of ocular preference is discussed first.

In the cat, binocular blockade of retinal activity between the second and fifth postnatal weeks via tetrodotoxin injections prevents the formation of ocular dominance columns in layer IVC (Stryker & Harris, 1986). In contrast, binocular lid suture or dark rearing from birth to six weeks or later permits a considerable amount of segregation to occur. When the two eyes are visually stimulated alternately between tetrodotoxin treatments, the segregation of geniculocortical afferents occurs normally. However, when the eyes are stimulated in synchrony the segregation is prevented (Stryker & Strickland, 1984). Additional findings suggest that any imbalance in activity between the two eyes is sufficient to cause a shift in the ocular dominance distribution in favor of the eye that is allowed to sustain a relatively higher amount of spontaneous discharge (Chapman et al., 1986). Thus, complete blockade of activity in one eye was sufficient to cause a small, yet significant, shift in ocular preference in favor of the other eye, even when the latter was either light- or pattern vision-deprived during the same period (between the fourth and the sixth postnatal weeks). In another type of experiment, developing animals are exposed daily to stroboscopic light which forces the afferents from the two eyes to fire synchronously. This treatment also prevents segregation, presumably by eliminating differences in the firing patterns between the afferents from the two eyes, thus preventing interocular competition among geniculocortical afferents. In sum, at the cellular level two basic conditions are required for eye-specific segregation: 1) the afferents that originate from the same eye must display highly correlated patterns of activity and, 2) the afferents that originate in different eyes must display asynchronous firing patterns. While the former condition promotes cooperation among same-eye afferents, the latter condition enables competition among afferents that originate in different eyes and converge onto common cortical targets.

In most mammals, including primates, the segregation of geniculocortical afferents begins before the onset of visual experience. Given the evidence presented above, correlated spontaneous activity among afferents appears to play an important role in the initial stages of eye-specific segregation in lamina IVC. Upon their arrival in layer IVC, the terminals of afferent neurons that are dominated by one eye are intermixed with terminals that carry inputs originating from the other eye. At that time, only spontaneous activity is present in these afferents. Moreover, neurons that innervate adjacent retinal loci in one eye show highly correlated patterns of spontaneous activity whereas little or no covariance in interocular firing patterns is found (e.g., Meister et al., 1991). In addition, random variations in the distribution of eye-specific terminals, guided by selective adhesion and chemoaffinity processes, give each eye a slight regional advantage over the afferents from the other eye in lamina IVC. It is very likely that the neurophysiological process that controls the formation of ocular dominance columns capitalizes on regional intercorrelations in spontaneous firing and random variations in the spatial distribution of axonal terminals. Regions that show these small imbalances in the distribution of terminal branches serve as "centers" for the emergence of ocular dominance columns. The dominance of single-eye afferents within these primitive ocular preference columns is subsequently strengthened by activity-dependent processes. Stabilization of the synapses formed by the "preferred" eye depends upon their relative advantage in driving their postsynaptic targets, ensured by superior synaptic density. The capacity of inputs originating from one eye to drive target neurons is further enhanced by the increased likelihood of synchronous (or highly intercorrelated) firing among geniculocortical afferents that receive input from adjacent retinal loci. Both spontaneous and visually induced retinal activity would take advantage of this process. Conversely, contacts formed by the other eye within the same region are less likely to become stabilized because they are less likely to drive their postsynaptic targets due to insufficient synaptic density. In addition, they cannot take advantage of postsynaptic firing induced by the afferents from the other eye because they do not fire in synchrony with those afferents.

An important component of synaptic stabilization that is governed by Hebbian rules, is that the postsynaptic cell must be driven by the presynaptic cell. In other words, a prerequisite of stabilization is a strong correlation between pre- and post-synaptic activity. Synchronous activity in the afferents that converge onto a single target cell ensures that the postsynaptic cell will be sufficiently depolarized. Several lines of evidence are consistent with this claim. For instance, local intracortical application of the GABA agonist muscimol (which selectively binds to the postsynaptic GABA$_A$ receptor) during monocular lid-suture in four to five week-old kittens results in a shift of ocular preference in favor of the closed eye (Reiter & Stryker, 1988). This surprising outcome was observed even after the immediate effects of muscimol had worn off, and was restricted to areas where electrophysiological activity was effectively blocked by the muscimol injections. Outside of these regions monocular deprivation was associated with a shift in ocular preference in favor of the nondeprived eye, as expected. This study provided the first direct demonstration that suppression of postsynaptic activity in the visual cortex can alter the outcome of monocular deprivation without affecting activity in the presynaptic geniculocortical afferents. These findings suggested for the first time that postsynaptic firing is a necessary element in mediating the effects of activity-dependent eye-specific segregation in the visual cortex.

There is also evidence at the cellular level that correlated pre- and post-synaptic activity may increase the transmission efficacy and facilitate the stabilization of already existing synapses in the visual cortex (Frégnac et al., 1988). In that study the postsynaptic firing of visual cortical cells was manipulated iontophoretically during monocular visual stimulation in kittens. Changes in the strength of ocular preference in favor of one eye occurred only when excitatory iontophoretic pulses were temporally correlated with stimulation of that eye. Approximately 30% of all the neurons examined in that study showed some shift in ocular preference in favor of the previously nondominant eye. This shift lasted for up to 100 min. following repeated iontophoretic pulses. In a few cells, the shift lasted for several hours. No such changes were observed in any of the following conditions: 1) iontophoretic induction of firing that was not temporally correlated with the onset of visual stimulation, 2) iontophoretic currents that reduced or completely blocked postsynaptic activity, or 3) in the absence of iontophoretic stimulation. Interestingly, similar changes were observed in adult cats. It is not known at present if the same cellular mechanism is responsible for synaptic modification phenomena in kittens and in adult cats.

The hippocampus is a prime model of synaptic enhancement phenomena. High-frequency (tetanic) electrical stimulation of the major afferent pathway to the hippocampus can lead to prolonged enhancement of synaptic activity in this structure (Kelso et al., 1986). This phenomenon is known as long-term potentiation (LTP). In the hippocampus the capacity of a weak synaptic signal to drive its postsynaptic target is increased by either pairing it with a strong input conveyed via another afferent, or by iontophoretic depolarization of the postsynaptic cell in synchrony with the primary, yet weak, afferent input (Bindman et al., 1988; Gustafsson et al., 1987). In both the hippocampus and in the neocortex, LTP requires strong postsynaptic depolarization, which can often be achieved only if the inhibitory action mediated by GABA receptors is simultaneously suppressed (Artola & Singer, 1987).

LTP-like phenomena have been demonstrated in the cat following tetanic stimulation of the optic nerve in vivo (Tsumoto & Suda, 1979). More recently, LTP was successfully induced in slices taken from visual cortex in the same species. Changes in synaptic transmission that lasted for up to 15 hours at a time were observed following prolonged low frequency stimulation of the white matter (with 15- to 60-minute trains of 2 Hz pulses; reviewed by Toyama et al., 1993). The likelihood of synaptic potentiation was highest in slices that were taken from kittens at the age of three to five weeks. This period corresponds to the window of peak susceptibility for the emergence of ocular dominance in the visual cortex (Smith, 1985).

In addition to LTP, long-term synaptic depression (LTD) has also been documented in the visual cortex of the cat in vivo and also in rat visual cortex slices. For instance, Tsumoto & Suda (1979) showed that unilateral low frequency stimulation of the optic nerve is capable of suppressing of synaptic transmission triggered by stimulation of the contralateral optic nerve. As in the case of LTP, susceptibility to long lasting synaptic depression parallels the course of the sensitive period for the formation of ocular dominance columns in layer IVC (Dudek & Friedlander, 1996). According to one view, LTD is partially responsible for the reduced capacity of inputs from the deprived eye, in monocularly deprived animals, to drive cortical neurons (Bear & Rittenhouse, 1999). Persistent failure to drive postsynaptic cells, not only prevents the onset of LTP phenomena, but eventually leads to further suppression of neuronal signaling at these synapses. The reduction in synaptic transmission is, according to this view, the result of an active neurochemical process that is initiated in the postsynaptic cell and, on the long run, could lead to physiological or even anatomical changes in the presynaptic axon.

Molecular mechanisms involved in activity-dependent phenomena

The role of neurotrophins in normal development and plasticity
Normal development of the visual cortex and phenomena of neuronal plasticity during development involve rearrangement in the projection patterns of afferent neurons (during early stages) and significant changes in patterns of axonal and dendritic branching within cortical layers (during later stages). These observations suggested a role for neurotrophins, which were known to promote neuronal growth in other parts of the central nervous system. There are at least four molecules that belong in the family of neurotrophins, each exerting its action through specific tyrosine kinase receptors: nerve growth factor

(NGF), brain-derived growth factor (BDNF), neurotrophin-3 (NT-3), and neurotrophin-4/neurotrophin-5 (NT4/5).

According to this notion the outcome of the competition between neurons that project onto common targets in the visual system ultimately determines the amount of neurotrophins available to sustain existing axonal branches and for promoting further growth. Trophic molecules released by the postsynaptic cell interact with the presynaptic cells and operate as retrograde[4] messengers to trigger protein synthesis, which is required for the structural changes mentioned above. For instance, competition between geniculocortical afferents, that project to the same layer IVC cells, for a limited supply of neurotrophins may determine which axons or axonal arbors will ultimately survive. Given that competitive interactions between neurons are guided by activity-dependent processes, researchers closely examine the role of neurotrophins as translators of neuronal signaling into long-term changes in patterns of connectivity. The evidence available to date in support of this proposal can be summarized as follows:

- Neurotrophins enhance neuronal growth in the visual cortex. For instance, BDNF promotes axonal branching of retinal ganglion cells in amphibians in vivo (Cohen-Cory & Fraser, 1995). Moreover, the effects of neurotrophins in regulating dendritic growth are layer-specific, with each type promoting a distinct growth pattern (McAllister et al., 1995). Finally, downregulation of endogenous neurotrophins has a significant impact on dendritic growth (McAllister, Katz, & Lo, 1997).
- Regulation of endogenous neutrotrophic molecules and their receptors follows a temporal course that is distinct for each visual cortex layer, and correlates with the course of normal development, including neuronal growth and synaptogenesis (Cabelli et al., 1996; Lein et al., 2000).
- Neurotrophins appear to play an important role in regulating normal development of functional neuronal properties in the visual cortex and in modulating the effects of visual deprivation: a) supply of antibodies against NGF during development arrests the maturation of receptive field properties and effectively prolongs the sensitive period for monocular deprivation (Pizzorusso et al., 1997); b) intraventricular or intracortical injection of NGF prevents the effects of monocular deprivation (Maffei et al., 1992) and dark rearing, such as the arrest in the maturation of orientation selectivity (Berardi et al., 1994).

Unfortunately, there is little information on the effects of visual experience in the regulation of neurotrophin levels in the visual cortex. There is some evidence that the action of neurotrophins depends upon neuronal activity. Thus, in hippocampal slices, intracellular levels of BDNF and NGF can increase rapidly in response to iontophoretic depolarization and decrease following activation of inhibitory (GABA$_A$) receptors (Lu et al., 1991; Zafra et al. 1991). Conversely, levels of the messenger RNA that transcribes BDNF in the visual cortex are reduced following blockade of retinal activity (Schoups et al., 1995). Following monocular activity blockade with TTX injections, downregulation of BDNF mRNA is restricted to cortical regions that presumably correspond to ocular dominance patches favoring the deprived eye (Lein & Shatz, 2000). Moreover, the effects of endogenous neurotrophins on dendritic growth require neuronal signaling (McAllister et al., 1996).

The viability of activity-dependent processes as regulators of normal neuronal development and developmental plasticity rests on the premise that specific patterns of neuronal activity can lead to long-term changes in synaptic efficacy. The most popular hypothesis is that correlated pre- and post-synaptic activity induces various forms of long-term synaptic potentiation (LTP), while "uncorrelated" presynaptic activity and postsynaptic firing is associated with long-term synaptic depression (LTD). Long term synaptic modification is subsequently "translated" into lasting changes in neuronal structure, which would include patterns of axonal branching, distribution of postsynaptic sites, etc. Although neurotrophins are prime candidates for the role of the "translator" in this process, a direct forward link between neurotrophin action and LTP or LTD phenomena has not been demonstrated in the visual cortex. There is also little evidence that modulation of neurotrophin action can indeed occur in a precise time-limited manner, and with the spatial specificity, at the subcellular level, required to mediate Hebbian-rule based phenomena in the developing visual cortex.

Molecular processes involved in long-term synaptic modification
In the visual cortex, long-lasting changes in synaptic efficacy appear to reside primarily in the action of the N-Methyl-D-Aspartate (NMDA) receptor. The unique features of this receptor (for a review, see Watkins, 1994), render it ideal for mediating phenomena of long-term synaptic enhancement involving the detection of correlated pre- and postsynaptic activity. There are several pieces of evidence linking the NMDA receptor with LTP phenomena. Thus, the induction of LTP requires and increase in the intracellular calcium concentration in the postsynaptic cell (Kimura et al., 1990). Indeed, one of the characteristic effects of NMDA receptor activation is calcium influx. As expected, LTP is abolished when the NMDA receptors are blocked pharmacologically (Artola & Singer, 1990). Moreover, activation of the NMDA receptor-ionophoric channel complex requires, in addition to the binding of an agonist molecule on the NMDA receptor itself, that the postsynaptic membrane becomes depolarized, probably via activation of a different receptor system.

Although there is an abundance of data that suggest a role of NMDA receptors in long-term synaptic enhancement effects in the hippocampus (for a recent review, see Bashir et al., 1994), evidence for a contribution of NMDA receptor-mediated LTP phenomena in developmental plasticity is still sparse. However, several findings are consistent with the proposed role of NMDA receptors in developmental plasticity in the visual cortex. First, NMDA receptors are found in abundance in the visual cortex, where they seem to make a strong contribution to the normal responsivity to light of cells in layers II and III to across a wide range of ages (Fox et al., 1989). Further, the contribution of NMDA-receptor activation to excitatory transmission appears to be greater early in development (Fox et al., 1989). Thus, the number of NMDA receptors and calcium channels is larger in the developing animal (Bode-Greuel & Singer, 1988, 1989). Second, infusion of the NMDA receptor blocker D-2-amino-5-phosphonovalerate (APV or AP5) mitigates the severity of the effects of monocular deprivation in the cat visual cortex (Kleinschmidt et al., 1987). In addition, APV prevents the restoration of ocular dominance preference following reverse suture treatment (Gu et al., 1989). These findings suggest that the NMDA receptor may be involved in the processes that mediate both the selective elimination of the afferents that experience a competitive disadvantage, as well as the reinforcement of synaptic contacts that acquire a relative advantage as a result of monocular deprivation.

Other mechanisms involved in the modulation of long-term synaptic modification in the visual cortex involve the action of two neurotransmitters, acetylocholine and norepinephrine, and endogenous neurotrophins. Intraventricular or intracortical application of 6-hydroxydopamine, an agent that causes depletion of norepinephrine, can prevent the expected ocular dominance shift following monocular deprivation imposed during the critical period (Kasamatsu & Pettigrew, 1976; Kasamatsu et al., 1979). Ocular dominance shifts can be prevented when both the norepinephrine and the cholinergic projection from the brainstem to the visual cortex is disrupted (Bear & Singer, 1986), suggesting that the two transmitters play a role in mediating regulatory brainstem influences on cortical maturation. At the cellular level, parallel action of noradrenergic and cholinergic systems can facilitate the induction of LTP (Bröcher et al., 1992). These effects could be exerted independently or indirectly by modulating NMDA-receptor action. For instance, it is possible that the two systems operate synergistically to increase postsynaptic depolarization and increase the likelihood that the NMDA-receptor gate calcium channels will open leading to a greater influx of calcium. Finally, the long-term maintenance of LTP in rat visual cortex slices requires the activation of a specific neurotrophin receptor (Sermasi et al., 2000). The purported role of neurotrophins, such as NGF and BDNF, in LTP phenomena may be related on their capacity to enhance glutamate release (Sala et al., 1998).

At present, there is only indirect evidence that cortical LTP is associated with structural synaptic modifications such as an increase in the number of postsynaptic dendritic spines (Perkins & Teyler, 1988). There are, however, a number of hypotheses regarding the mechanisms involved in the process of translating prolonged changes in synaptic efficacy into relatively permanent synaptic properties. With respect to LTP, it is believed that an important role in this process is played by the elevated concentration of intracellular calcium caused by the opening of NMDA receptor-channels (reviewed by Tsumoto, 1993). This process had been originally implicated in LTP phenomena in the hippocampus. Intracellular calcium can trigger several second messenger systems, including cyclic adenosine monophosphate (cAMP), and a variety of protein kinases, such as the calcium/calmodulin-dependent protein kinase II (CaMK II), protein kinase A, and protein kinase C. Second messengers are capable of mobilizing a number of immediate early genes (IEGs), such as the cAMP responsive element (CREB; Bourtchuladze et al., 1994). IEGs, in turn, promote transcription of other genes that trigger protein synthesis (directly or indirectly through intracellular regulation of neurotrophins; Tao et al., 1998) required for long term changes in synaptic connections (Sheng & Greenberg, 1990).

Second messenger systems are indeed activated by neuronal activity and visual experience. It has been shown that in layers II and III of the visual cortex of the rat the presence of CaM II determined whether tetanic stimulation of the white matter would elicit LTD- or LTP-like effects. Increased levels of CaM II have been found in the ocular preference columns dominated by the deprived eye as early as two days following monocular deprivation in young primates. There is also evidence that the expression of protein kinase C in the visual cortex depends upon afferent input (reviewed by Tsumoto, 1993, p.574-5).

Finally, there are indications that LTP is associated with changes in the *presynaptic* terminals, in addition to the postsynaptic effects mentioned above. At least in the hippocampus, it was found that LTP is involved in the phosphorylation of protein GAP-43 which, in turn, interacts with cytoskeletal proteins and synaptic vesicles to control transmitter release (Norden et al., 1991). This is consistent with reports of increased transmitter release from presynaptic terminals during LTP in the hippocampus (Bekkers & Stevens, 1990). A protein analogous to GAP-43 in the goldfish is involved in the sharpening of the retinotopic map that occurs following resection and regeneration of the optic nerve (Benowitz & Schmidt, 1987).

There have been several attempts to explain how LTP phenomena, which are initially expressed by postsynaptic changes, can lead to persistent modifications in the presynaptic terminal. A common element in these proposals is that LTP is associated with the production of certain chemical agents in the postsynaptic membrane. Two of the hypothetical mechanisms are briefly mentioned here. One possibility is that there is an increase in the amount of intercellular adhesion molecules in the postsynaptic membrane, and this leads to an increase in the strength of the attachment between the cells and/or the size of the synaptic contact itself (Desmond & Levy, 1986). According to a second proposal, a diffusible agent released from the postsynaptic membrane binds onto specific receptor sites in the presynaptic membrane and triggers a chain of events that lead to increases in transmitter release and, perhaps, also to structural changes in the presynaptic terminal (Williams & Bliss, 1988). Two compounds that have received extensive attention as potential diffusible agents are the arachidonic acid and nitric oxide (Cramer et al., 1996). It should be noted that the hypothetical mechanisms described above are not mutually exclusive. Moreover, their involvement in long-term synaptic modification has not yet been conclusively demonstrated neither in the hippocampus nor in the visual system.

The role of normal binocular vision in the emergence of binocular neurons in the visual cortex and the development of stereoscopic depth perception

As mentioned in the previous chapter, the emergence of neurons with binocular response properties in the non-afferent layers of the primary visual area requires that the eye-specific segregation of geniculocortical afferents arriving in layer IVC has reached an advanced stage. Before discussing the presumed role of experience in the development of cortical binocularity, essential information regarding the normal development of binocularity in humans is presented. Non-invasive electrophysiological recordings in the form of visual evoked responses fail to show evidence for either cortical binocularity or stereopsis in infants younger than three months. The majority of infants older than four months of age show evidence of cortical binocularity when tested with dynamic random dot correlograms (Braddick et al., 1980; Petrig et al., 1981), and with random dot stereograms (Petrig et al., 1981). The data from these studies provide strong evidence for the presence of binocular neurons in the visual cortex of young infants. However, the presence of binocular cortical neurons does not necessarily imply the existence of disparity sensitive mechanisms (Fox, 1981; Shea et al., 1987). Shea and Dumais (1981) examined the ability of young infants (2.5 to 4.5 months of age) to perceive a stereoscopic form created by varying the degree of horizontal retinal disparity of corresponding elements in the center of a dynamic random-dot display. The displays used in the study did not contain monocular form or depth cues, and the rectangular target could only be seen when retinal disparity was introduced (i.e., when the subject was viewing the display through different chromatic filters placed before each eye). The results indicated that infants aged between 3.5 and 4.5 months detected the moving stereoscopic form whereas 2.5 month-olds as a group did not. These findings are in close agreement with data obtained using traditional line stereograms (Birch et al., 1985; Held et al., 1980; Shimojo et al., 1986). Overall, these data place the onset of functional stereopsis around 3.5 to 4 months of age for the majority of infants with normal developmental history.

Taken together, behavioral and electrophysiological studies lead to the following conclusions: 1) cortical binocularity and stereopsis are poorly developed during the first 2 to 3 months of life; their level of maturation is not sufficient to produce either synchronized evoked activity recorded from the scalp or reliable behavioral orienting toward a stereoscopic form, 2) cortical binocularity and stereopsis appear to emerge in close temporal synchrony between the ages of 3 and 4 months in the majority of infants with normal developmental histories.[5]

Despite the wealth of information produced by studies with human infants, the conditions necessary for, and the processes involved in, the development of cortical binocularity cannot be studied directly in humans. In an attempt to answer such questions, research turned to animal models that permit the use of invasive experimental manipulations. Two main conditions are traditionally linked with the development of cortical binocularity: 1) the relative amount and quality of patterned visual input to the two eyes and, 2) alignment of the two lines of sight. The former condition can be studied experimentally through a variety of monocular deprivation protocols, whereas the latter can be studied by observing the effects of experimentally induced strabismus.[6] In primates, an abundance of binocular neurons can be found in the non-afferent layers of the primary visual cortex at birth (Wiesel & Hubel, 1974). Short periods of monocular deprivation during the first three weeks can prevent the emergence of binocular neurons in the upper layers almost completely (LeVay et al., 1980). Equally short periods of monocular deprivation, between three and eight weeks, can result in what appears to be a permanent loss of binocular neurons in area V1 (Crawford & von Noorden, 1980; LeVay et al., 1980) and V2 (Crawford et al., 1984). Although the number of binocular cells reaches a peak around the sixth week (LeVay et al., 1980), disruption of normal binocular vision imposed as late as the end of the first year may still affect cortical binocularity, provided that it lasts long enough (Blakemore et al., 1978; LeVay et al., 1980).

Another experimental intervention, used to disrupting the balance of input from the two eyes to the brain, consists of repeated, asynchronous electrical stimulation of the two optic nerves. In these experiments, normal visual input from both retinas is suppressed through intraocular infusion of tetrodotoxin. Electrical stimulation of the optic nerves in kittens, lasting for one-to-two hours every day prevents the emergence of all but few binocular neurons in upper layers of V1 (Stryker & Strickland, 1984). As a rule, susceptibility to disruptions of normal binocular vision persists for a longer period in the non-afferent layers than susceptibility of eye-specific segregation in layer IVC. In addition to disruptions of balanced binocular input, misalignment of the lines of sight associated with experimentally induced strabismus also affects the development of cortical binocularity (Crawford & von Noorden, 1980). This intervention can both prevent the emergence of binocular neurons in V1 and cause dramatic reductions in the proportion of binocular neurons in older, yet still, developing animals. To summarize, balanced input from the two eyes is necessary not only for the development of the normal, adult-like pattern

of ocular dominance columns in layer IVC, but also for the establishment and maintenance of vertical projections from the afferent layers that converge onto common targets
in the nonafferent layers. However, in contrast to the process of eye-specific segregation,
the development of cortical binocularity requires *synchronous*, balanced input from the
two eyes.

The notion that interocularly correlated input plays a role in the emergence of normal
binocular properties in the visual cortex is consistent with the proposed role of Hebbian
rules in this process (Constantine-Patton et al., 1990). At the macroscopic level, the hypothesized mechanism is essentially similar to the one postulated to explain activity-
dependent ocular segregation in layer IVC, as discussed above. Binocular neuronal properties are believed to mature along with improvement of binocular alignment, which
makes bifoveal vision possible and also improves the likelihood of correlated firing patterns among afferents that carry input from adjacent retinal loci and converge upon
common target cells in non-afferent layers. At the cellular level, the NMDA receptor has
been implicated in the development of binocular convergence by mediating the effects of
correlated neuronal activity. Indeed, there is preliminary evidence that an NMDA-
receptor mediated process may be responsible for the plasticity of binocular tectal maps
in the Xenopus frog, following $90°$ rotation of one eye (Schrerer & Udin, 1991).

Binocular cells in the upper layers of the primary visual cortex are important for perceptual capacities that rely on binocular vision such as stereopsis. Indeed, one may argue
that the biological significance of binocularity is primarily to support stereopsis (Shinkman et al., 1985). Primates raised with experimentally induced strabismus, which prevents the emergence of binocular neurons in areas V1 and V2, appear to lack stereoscopic vision when tested with random dot stereograms (Crawford et al., 1984). Deprivation studies in kittens indicate that manipulations, which preserve a balanced amount of
form vision to each eye separately, but prevent simultaneous stimulation of both eyes
with visual contours,[7] can also prevent the emergence of binocular neurons in the primary visual cortex. In addition, behaviorally assessed stereoscopic vision is found to be
severely impaired in these animals (Blake & Hirsch, 1975).

On the other hand, cortical binocularity may not be sufficient to maintain stereoscopic
vision. For instance, dark-rearing in kittens results in loss of stereopsis despite the presence of a substantial number of binocular neurons in striate cortex. A similar situation
may also occur in humans: using measures of interocular transfer of the tilt after-effect, a
number of investigations have shown that binocular processes may be present in individuals who appear completely stereoblind (Aslin, 1981). Another example of a dissociation between binocularity and stereopsis is described below.

In addition to horizontal disparity, a binocular cue that can be used to support stereopsis is orientation disparity. In the cat, many binocular neurons are tuned to a narrow
range of orientation disparities. Certain rearing conditions have been shown to affect
neuronal sensitivity to orientation disparity, without impairing cortical binocularity, per
se. In a series of experiments, Shinkman and colleagues (reviewed in Shinkman et al.,
1985) raised kittens, from their fourth week to one or two months of age, with prisms that
induced different degrees of interocular orientation disparity. One group of kittens were
daily exposed to $16°$ orientation disparity, a value that is near the maximum disparity
values observed among binocular cells in normally reared animals. Two other groups
were exposed to disparities that were clearly beyond the normal upper disparity limit for

binocular cells (24° and 32°, respectively), while a fourth group wore goggles that induced zero disparity (control group). The mean preferred orientation disparity among visual cortical cells in the 16°-prism group was shifted (compared to the control group) by an amount equal to the prism-induced orientation disparity during rearing. The majority of cells in the 16° condition were binocular, and their orientation disparity preferences were tightly clustered around the mean, which in this case was approximately 16°. Further, there was a relatively orderly succession of orientation disparity preferences along oblique electrode penetrations. Animals exposed to larger orientation disparities during rearing had very few binocular neurons. These findings are consistent with the possibility that tuning of individual cells in the visual cortex to a narrow range of retinal disparity values can be modified under certain conditions of visual stimulation. These findings are consistent with the "instructional" hypothesis, in that the stimulus selectivity of individual neurons can be altered by particular types of visual experience. The outcomes from the two groups reared under extreme orientation disparity conditions were consistent with the predictions of the "selective" hypothesis. The data showed that the only neurons that maintained some degree of selectivity were those with an initial bias toward the extreme orientation disparities that the animal was exposed to during rearing.

As mentioned in the beginning of this section, binocularity emerges between three and four months in the majority of human infants with normal developmental histories (Birch et al., 1985; Fox et al., 1981; Held et al., 1980). In primates, the *susceptibility* of binocularity to disruptions of binocular vision, due to misalignment of the eyes, often extends throughout the first year of life (Blakemore et al., 1978; LeVay et al., 1980). Based on these findings, one could predict that the sensitive period for cortical binocularity in the human infant extends from three to four months to some time during the third year of life. Data from Banks et al. (1975) essentially confirms this prediction. In that study, binocularity was assessed by performance on the interocular transfer of the tilt after-effect in 24 individuals diagnosed with nonparalytic esotropia, with onset either during early infancy (Early onset group) or during late infancy and childhood (Late onset group). Estimates of the amount of normal binocular experience were obtained from each subject by taking into account the time of onset and the time when corrective surgery had been performed. This estimate was then adjusted in an iterative fashion in order to maximize the correlation between normal binocular experience and scores on the test for binocularity. This was done to optimize estimates of the relative importance of particular developmental periods for the maturation of binocularity and stereopsis. In general, the data placed the onset of the sensitive period in the first few postnatal months and its peak between 1 and 3 years of age. These estimates fit well with the suggested course of the sensitive period in primates, when a correction factor of 4:1 or 5:1 is applied (Boothe et al., 1985). To summarize, the time at which an advanced level of phenotypic development is reached (i.e., around 3-4 months in humans) coincides with the onset of the sensitive period, during which disrupting specific aspects of visual experience (in this case binocular vision) may permanently alter this phenotype.

The role of binocular pattern vision in the development of spatial resolution: behavioral and neural correlates

The development of normal pattern vision, as indexed by the two related measures, acuity and contrast sensitivity, depends upon the same general conditions required for the emergence of cortical binocularity: balanced patterned input from the two eyes and alignment of the lines of sight. Reduced spatial resolution in one eye caused by experimentally induced deprivation of form vision is known as *deprivation amblyopia*, whereas impairments in spatial resolution associated with ocular misalignment is known as *strabismic amblyopia*. As we will see below. different neurophysiological mechanisms appear to be primarily responsible for each type of amblyopia. First, we will discuss the case of deprivation amblyopia.

Depending upon the timing of the monocular deprivation treatment, spatial resolution in the deprived eye can be restored to a certain extent, following eye-opening. There is evidence that the extent of recovery is greater when the deprived eye is given a relative advantage over the previously non-deprived eye. One way to manipulate the balance of inputs from the two eyes is to disrupt vision in the previously non-deprived eye, a treatment known as *reverse deprivation*. In a series of experiments, Smith and Holdefer (1985) subjected kittens that were monocularly deprived for periods ranging from 27 to 50 weeks starting at the time of eye opening to one of three treatments: 1) enucleation of the nondeprived eye after a period of normal binocular experience that ranged between three days and 64 weeks (DE group), 2) lid suture of the experienced eye following a period of binocular vision ranging from three days to 22 weeks (RS group) and, 3) unobstructed binocular vision until the time of testing (i.e., for periods ranging between 24 and 41 weeks; BE group). Following the experimental treatment, psychophysical acuity curves were obtained for the initially deprived eye, from all animals, using the jumping stand apparatus. The animals that had not been subjected to reverse deprivation (BE group) showed the lowest acuity estimates among the three groups. Overall, animals in the DE group showed significantly higher spatial resolution than the animals in the RS group. In conclusion, acuity in the eye that was deprived of form vision for the entire extent of the critical period appears to correlate with the amount of competitive advantage given to that eye over the initially non-deprived eye.

Neurophysiological data obtained from kittens and primates with demonstrated amblyopia agree with respect to two main findings: 1) the proportion of V1 neurons that respond to visual stimulation of the deprived eye is significantly lower than the proportion of cells that respond to the non-deprived eye; moreover, a strong linear relation between the number of cortical neurons driven by the deprived eye and acuity estimates for that eye has been reported (Smith, 1985) and, 2) neurons that remain responsive to the deprived eye have diffuse and significantly less sensitive receptive fields; as a result their contrast sensitivity is significantly reduced (Movshon et al., 1987). However, extended periods of reverse-suture in kittens, that leads to a significant improvement in visual acuity, is typically associated with a dramatic increase in the number of neurons that respond to the deprived eye, but not necessarily to an improvement in the spatial resolution of these neurons (Smith et al., 1978). This finding suggests that an important correlate of the observed improvement in spatial resolution of the previously deprived eye is its ability to drive a substantial proportion of cortical cells. It is therefore possible that in monocularly deprived animals (where the spatial resolution of individual neurons is impaired) visual acuity is determined by the density of the receptive fields of visually responsive cells. Fewer responsive cells may lead to a reduction in the sampling density within the visual field causing a reduction in visual acuity. In conclusion, whereas in visually deprived animals the limits of spatial resolution in the visual system is determined by the sampling density of functional afferent neurons, in non-deprived animals it is the best spatial resolution of individual neurons that sets the limit for visual acuity (e.g., Blakemore & Vital-Durand, 1986, see also Chapter 4).

In humans, monocular visual form deprivation can be caused by a number of congenital conditions such as blepharoptosis, corneal opacity, severe anisometropia, uncorrected aphakia, and catarax. If the disruption of pattern vision is severe, deprivation amblyopia develops, which is considered as the most severe form of amblyopia, marked by a large reduction of visual acuity in the deprived eye, which is often irreversible (von Noorden, 1981). An evaluation of visual acuity before treatment in 11 cases of deprivation amblyopia with early onset due to a variety of causes revealed that spatial resolution can become severely impaired with as little as 1.5 months of monocular form deprivation that occurs during the first 4.5 years (von Noorden, 1981). Thus, it appears that amblyopia can still develop after the age at which acuity reaches adult levels. Normal acuity may develop if the ocular abnormality is corrected within the first few months of life. Infants who suffer from bilateral congenital catarax, which are corrected before the age of 4 to 6 months, may develop normal visual acuity by the end of their first year (Jacobson et al., 1981). A similar age window has been reported for the recovery of spatial resolution in cases of unilateral catarax. Surgical treatment of monocular catarax can be optimized by part-time occlusion of the intact eye (reverse deprivation).

In addition to monocular deprivation, deficits in spatial resolution can be induced by consistent misalignment of the lines of sight *(strabismic amblyopia)*. This condition is particularly interesting because, in contrast to deprivation amblyopia, each retina receives a balanced amount of visual stimulation. Indeed, it appears that the deviating eye is transmitting a sharp, well-focused image to the brain. Although the neurophysiological processes responsible for strabismic amblyopia are not completely understood, they appear to be different than those responsible for the development of deprivation amblyopia. Thus, studies on strabismic cats and primates have failed to find a significant reduction in the proportion of cells responsive to the deviating eye (Blakemore & Vital-Durand, 1992; Crewther & Crewther, 1990). In addition, the effects of early strabismus on the spatial resolution of individual cortical neurons is, at best, modest (Blakemore & Vital-Durand, 1992). Rearing kittens with surgically induced unilateral, convergent or divergent, strabismus until the age of four years does not seem to reduce the proportion of orientation-nonselective cells driven by the deviating eye within the cortical representation of the central visual field (Kalil et al., 1984). It appears, then, that neither reduced neural sampling of the visual image nor impaired spatial resolution of individual neurons is primarily responsible for the often dramatic impairment in visual acuity and contrast sensitivity found in the deviating eye of animals raised with experimentally induced strabismus. On the other hand, there is evidence, although largely circumstantial at present, that abnormal binocular interactions at the cortical level are responsible for the development of strabismic amblyopia. There have been reports suggesting reduced binocular facilitation as well as interocular suppression in V1 neurons in the cat and primates (Sengpiel et al.,

1994; Sengpiel & Blakemore, 1996). These findings are in agreement with evidence of reduced synchronized firing among neurons with similar orientation preferences that are located in neighboring columns dominated by different eyes (Roelfsema et al., 1994). It has been proposed that aberrant functional connections among neurons with similar visual pattern selectivity, but different ocular preferences, may underlie the phenomenon of interocular *suppression* of vision in the deviating eye that is typically observed after prolonged periods of significant misalignment of the eyes. The significance of interocular suppression, which appears to involve the persistence of predominantly inhibitory interactions between neurons that prefer one or the other eye, is consistent with the effects of reverse deprivation. Thus, suturing the non-deviating eye, at about the same time that misalignment is induced in the other eye, can prevent significant deterioration of spatial resolution in the deviating eye.

In conclusion, deprivation amblyopia can develop in animals and humans well after the time that visual acuity reaches adult levels. This finding suggests that patterned visual experience is necessary not only for the emergence but also for the *maintenance* of key aspects of the neuronal mechanisms responsible for normal spatial resolution.

Conclusion: The adaptive significance of developmental plasticity

In this chapter we have selectively reviewed the major milestones in the development of visual system properties that have been studied systematically in the past 30 years. To summarize, a number of developmental processes, such as cell division, neuronal differentiation, and migration of postmitotic cells into their final positions in the different parts of the visual system, are determined by a genetic code. At a later stage neurons in each structure send out afferent processes toward target structures, such as the tectum, dLGN, and primary visual cortex. In certain well-studied preparations, such as the growth of retinotectal projections in fish and amphibians, axonal pathfinding is determined, to a great extend, by various chemoaffinity agents. It is possible that the precision of certain developmental processes, such as the formation of intra- and inter-laminar connections in the visual cortex, depends upon chemical interactions between ingrowing axonal terminals and target cells.

At a latter stage, afferent axons form synapses with target cells, a large proportion of which are transient, characteristic of a distinctly immature pattern of synaptic connections. For instance, in the final days of gestation, in primates, the terminals of the geniculocortical afferents from the two eyes are almost entirely intermixed within layer IVC of the visual cortex in primates. Generally speaking, the segregation of geniculate afferents is mediated by two processes: the *cooperation* between the afferents that originate in the same eye, and the *competition* between the afferents that carry inputs from different eyes. At the cellular level, these processes depend upon the amount of neuronal signaling among presynaptic axons and their postsynaptic targets. Specifically, synaptic maturation depends upon 1) the increased capacity of converging afferents that carry inputs from the same eye to drive cortical target cells with which they synapse (cooperation), and 2) the reduced capacity of afferents from the other eye to drive the same cortical neurons (competition). At the molecular level, there is rapidly accumulating evidence that the effects of correlated pre- and post-synaptic activity are largely mediated through the action of NMDA receptors located in the postsynaptic cell membrane. These receptors are sensitive to both electrical and chemical events: that is, changes in the membrane potential as well as the binding of specific neurotransmitter molecules. This property renders NMDA receptors ideal for taking advantage of coincident activity in the pre- and post-synaptic cells. Activation of the NMDA receptor is associated with the mobilization of intracellular enzymes which, in turn, can induce persistent changes in the sensitivity of postsynaptic receptors, leading to increased synaptic efficacy.

Going a step beyond the simple description of processes and hypothetical mechanisms involved in the maturation of the visual system, a fundamental issue that remains to be addressed is the significance of plasticity for the development of the visual system and for vision. According to a widely accepted view, developmental plasticity enables the visual system to become "instructed" by specific external input. This argument is consistent with the observation that, at the onset of sensory experience, the central nervous system contains an excess of neurons and synapses, and that specific aspects of sensory stimulation regulate the selection of the most viable connections (Rauschecker, 1991). This view is based on two premises: 1) that perception in complex organisms depends upon the analysis of sensory cues that are far too numerous and complex to be coded in the genome, and 2) that is not feasible to "anticipate", by genetically encoding and transmitting, the full range of possible sensorimotor interactions between the organism and its environment. Under these constraints, one can easily foresee the adaptive value of a brain that shows a redundancy of cells and synapses at birth. The excess of neural elements establishes a substrate that is ready to receive and be shaped by a wide variety of afferent input. However, under normal circumstances, the rearing environment for individual members of a species contains a relatively stable range of stimuli that can be used to fine-tune and integrate sensory and motor systems. This relative stability, which has persisted throughout the evolutionary history of each species, has permitted the emergence of mechanisms that take advantage of specific patterns of afferent input (Greenough, 1985).

It appears that the impact of specific visual experience is regulated by two factors: 1) the developmental history of the individual and, 2) a central gating process. It shown that rearing animals in the dark delays the onset of the sensitive period for the emergence of neuronal selectivity to specific aspects of visual input (such as stimulus orientation). In addition, it is possible that the onset and extent of the sensitive period may be regulated by a central gating mechanism, which is in turn triggered by visual stimulation, not necessarily in a specific manner. According to one proposal, these regulatory signals may be controlled by intralaminar thalamic nuclei. The impact of the latter during visual stimulation may depend upon the behavioral state of the animal and attention (Singer, 1985).

Next, we discuss two additional instances where plasticity may play a role in development. Although the examples may not directly relate to plasticity in the retinocortical pathway, they demonstrate the potential utility of plasticity in compensating for anatomical and physiological limitations of the developing organism. Consider, for instance, the process of "calibration" of the physiological mechanism that determines the precise relation between saccadic magnitude and the duration of the neural pulse generated by neuronal groups in the pontine reticular formation and rostral midbrain (Sparks, 1986). The duration of this pulse, which serves as input to the oculomotor nuclei, is proportional to the amplitude of the resulting horizontal or vertical saccade. Although the relation between pulse duration and saccadic magnitude remains relatively plastic during adulthood, the efficiency of the calibration process is particularly critical during early infancy. As postulated by Aslin (1988), it is likely that the relation between the retinal locus of a visual target and the amplitude of the saccade necessary to foveate that target, must change during early infancy in order to compensate for the migration of photoreceptors toward the central fovea (Yuodelis & Hendrickson, 1986). As a result of this process, the mapping of retinal loci onto specific locations within the visual field may not be constant during the early stages of development. If this is true, then the mechanism that controls saccadic amplitude must remain highly modifiable in order to accommodate the changes in the mapping of retinotopic onto visuotopic coordinates.

Another domain of visuomotor function where developmental plasticity appears to play a critical role is the "recalibration" of the physiological mechanism that controls changes in vergence in response to retinal disparity cues. A key to the development of sensitivity to retinal disparity is the establishment of retinal correspondence that is, the spatial matching of monocular receptive fields that provide inputs to binocular cortical neurons. In turn, retinal correspondence depends upon, and is limited by, epigenetic factors such as the growth of the eye and interpupillary distance (Singer, 1985). For instance, the distance between the two eyes increases by approximately 60% between birth and adulthood. The problem posed by development is how infants retain an apparently stable interpretation of retinal disparity cues despite a constantly changing relationship between the absolute magnitude of retinal disparity and the relative distance of objects in depth. If the correspondence between retinal disparity and perceived depth could not be constantly adjusted during development, infants' attempts to focus their eyes on a visual target that moves in depth in front of them would be consistently inaccurate (for a more detailed discussion of these issues see Banks, 1993). It follows that the system that mediates stereopsis must be capable of continuously readjusting the manner in which retinal disparity cues are interpreted, in order to ensure accurate bifoveal fixation.

NOTES

[1] These studies were conducted in the ferret, a species born with a considerably less mature visual system than the cat or monkey, offering researchers the opportunity to examine the effects of experimental manipulations before the emergence of orientation preference and iso-orientation maps in the visual cortex. The earliest age at which visually driven responses can be detected in the visual cortex in this species is postnatal day 21. Fully mature iso-orientation maps can be found using optical imaging around postnatal day 50.

[2] The contrast sensitivity function of the "ideal" observer represents an estimate of optimal performance, given the known anatomical and physiological constraints of the *peripheral* visual system at a given stage of development. The hypothetical performance of the ideal observer can be used as an index of the amount and quality of visual information available to the CNS. The difference in performance between the ideal newborn observer and the ideal adult reflects the contribution of receptoral factors in the reduced acuity and contrast sensitivity of the newborn infant.

[3] Different functional properties of the visual cortex may display different sensitive periods. For instance, maturation of binocular response properties in non-afferent layer neurons starts later than the peak of the sensitive period for the development of ocular dominance in afferent layers. In contrast, the sensitive periods for the maturation of ocular dominance, and for the development of orientation selectivity in cortical neurons, are largely overlapping. The extent of the sensitive period also depends upon the past history of the animal. Thus, dark rearing may considerably extend the sensitive period for ocular dominance and orientation selectivity (Mower et al., 1981).

[4] It is possible that neurotrophins act as anterograde messengers as well: for instance uptake of neurotrophin molecules by retinal ganglion cells and transport along retinogeniculate axons reported.

[5] Alternatively, stereopsis may actually lag behind the emergence of binocularity, but the infants' performance on tests of stereopsis is hindered by attentional and motivational deficits and poor alignment of their lines of sight.

[6] Strabismus is a congenital disorder manifested as a deviation of the lines of sight, usually in the horizontal plane (Von Noorden, 1980), and can be either *unilateral* or *alternating*. In the former case, the patient uses one eye consistently to focus on visual targets, while the other eye deviates. In alternating strabismus the patient fixates with one eye at certain times and with the other eye at other times. When the deviation is toward the midline the strabismus is called *esotropic*, whereas a lateral deviation is characteristic of exotropic strabismus or *exotropia*. The former is the most frequent type of strabismus. A large variety of clinical conditions can cause strabismus. By far the most frequent cause is hypermetropia (*accommodative strabismus*). Other causes include congenital deviation of both eyes, anomalies in convergence, and paresis of one or more of the extraocular muscles. In the periphery, the primary effect of strabismus is that visual images do not fall on corresponding retinal loci in the two eyes. Perceptually, this condition leads to diplopia (double vision) by preventing fusion of the images transmitted by each eye to the brain. The visual system adapts to this condition and maintains single vision by: 1) shifting retinal correspondence to match the angle of strabismic deviation; this anomalous retinal correspondence is produced by a yet unknown cortical process, or 2) interocular *suppression* of vision in the deviating eye. Suppression is believed to be one of the underlying causes of the reduction in spatial resolution in the non-fixating eye (strabismic amblyopia). Strabismus can be induced experimentally in animals, either surgically or by means of prismatic lenses worn in one or both eyes. The most common procedure for inducing strabismus is resection of the tendons of one or more extraocular muscles: resection of the medial rectus is usually sufficient to cause exotropia in the operated eye, whereas section of the lateral rectus is employed to induce esotropia.

[7] Such conditions can be created by using an *alternating monocular deprivation* protocol.

References

Allman, J.M., Meizin, F., & McGuinness, E. (1985). Direction and velocity-specific responses from beyond the classical receptive field in the middle temporal visual area (MT). *Perception, 14*, 105-126.

Andersen, R.A. (1989). Visual and eye movement functions of the posterior parietal cortex. *Annual Review of Neuroscience, 12*, 377-403.

Andersen, R.A., & Siegel, R.M. (1986). Two- and three-dimensional structure from motion sensitivity in monkeys and humans. *Society of Neuroscience Abstracts, 12*, 1183.

Andersen, R.A., Snowden, R.J., Treue, S. & Graziano, M. (1990). Hierarchical processing of motion in the visual cortex of monkey. *Cold Spring Harbor Symposia on Quantitative Biology, 55*, 741-8.

Antonini, A. & Stryker, M.P. (1993). Rapid remodeling of axonal arbors in the visual cortex. *Science, 260*, 1819-21.

Artola, A. & Singer, W. (1987). Long-term potentiation and NMDA receptors in rat visual cortex. *Nature, 330*, 649-52.

Aslin, R.N. (1981). Development of smooth pursuit in human infants. In D.F. Fisher, R.A. Monty, & J.W. Senders (Eds). *Eye movements: Cognition and visual perception* (pp. 31-51). Hillsdale, NJ: Erlbaum.

Aslin, R.N. (1988). Anatomical constraints on oculomotor development: Implications for infant perception.

Baizer, J.S., Ungerleider, L.G., & Desimone, R. (1991). Organization of visual inputs to the inferior temporal and posterior parietal cortex in macaques. *Journal of Neuroscience, 11*, 168-190.

Banks, M.S. (1980). The development of visual accommodation during early infancy. *Child Development, 51*, 646-66.

Banks, M.S., Aslin, R.N., & Letson, R.D. (1975). Sensitive period for the development of human binocular vision. *Science, 190*, 675-677.

Banks, M.S. & Bennett, P.J. (1988). Optical and photoreceptor immaturities limit the spatial and chromatic vision of human neonates. *Journal of the Optical Society of America, 5*, 2059-79.

Banks, M.S. & Shannon, E. (1993). Spatial and chromatic visual efficiency in human neonates. In C.E. Granrud (Ed.), *Visual perception and cognition in infancy*. Hillsdale: Lawrence Erlbaum, pp. 1-46.

Barlow, H.B. (1972). Single units and sensation: a neuron doctrine for perceptual psychology. *Perception, 1*, 371-394.

Bashir, Z.I., Berretta, N., Bortolotto, Z.A., Clark, K., Davies, C.H., Freguelli, B.G., Harvey, J., Potier, B., & Collingridge, G.L. (1994). NMDA receptors and long-term potentiation in the hippocampus. In G. L. Collingridge & J. C. Watkins (Eds.), *The NMDA receptor*, 2nd Edition. Oxford: Oxford University Press, pp. 294-312.

Bauer, R.M. (1993). Agnosia. In K.M. Heilman and E. Valenstein (Eds.), *Clinical Neuropsychology, 3rd Edition*. Oxford: Oxford University Press.

Baylis, G.C., Rolls, E.T., & Leonard, C.M. (1985). Selectivity between faces in the responses of a population of neurons in the cortex of the superior temporal sulcus of the monkey. *Brain Research, 342*, 91-102.

Baylor, D. (1992). Transduction in retinal photoreceptor cells. *Sensory Transduction;* The Rockefeller University Press: 151-174.

Baylor, D.A., Fuortes, M.G.F., & O'Bryan, P.M. (1971). Receptive fields of cones in the retina of the turtle. *Journal of Physiology, 214*, 265-294.

Bear, M.F. & Rittenhouse, C.D. (1999). Molecular basis for induction of ocular dominance plasticity. *Journal of Neurobiology, 41*, 83-91.

Bear, M.F. & Singer, W. (1986). Modulation of visual cortical plasticity by acetylocholine and noradrenaline. *Nature, 320*, 172-176.

Bekkers, J.M. & Stevens, C.F. (1990). Presynaptic mechanism for long-term potentiation in the hippocampus. *Nature, 346*, 724-728.

Benowitz, L.I. & Schmidt, J.T. (1987). Activity-dependent sharpening of the regenerating retinotectal projection in the goldfish: Relationship to the expression of growth associated proteins. *Brain Research, 417*, 118-121.

Berardi, N., Cellerino, A., Domenici, L., Fagioini, M., Pizzorusso, T., Cattaneo, A. & Maffei, L. (1994). Monoclonal antibodies to nerve growth factor affect the postnatal development of the visual system. *Proceedings of the National Academy of Science USA, 91*, 684-688.

Berry, M.J., Warland, D.K., & Meister, M. (1997). The structure and precision of retinal spike trains. *Proceedings of the National Academy of Science USA, 94*, 5411-5416.

Bindman, L.J., Murphy, K.P. & Pockett, S. (1988). Postsynaptic control of the induction of long-term changes in efficacy of transmission at neocortical synapses in slices of rat brain. *Journal of Neurophysiology, 60*, 1053-65.

Birch, E.E., Shimojo, S. & Held, R. (1985). Preferential-looking assessment of fusion and stereopsis in infants aged 1-6 months. *Investigative Ophthalmology & Visual Science, 26*, 366-370.

Blake, R. & Hirsch, H.V.B. (1975). Deficits in binocular depth perception in cats after alternating monocular deprivation. *Science, 190*, 1114-1116.

Blakemore, C., Garey, L.J., & Vital-Durand, F. (1978). The physiological effects of monocular deprivation and their reversal in the monkey's visual cortex. *Journal of Physiology, 283*, 223-62.

Blakemore, C. & Van Sluyters, R.C. (1975). Innate and environmental factors in the development of the kitten's visual cortex. *Journal of Physiology, 248*, 663-716.

Blakemore, C. & Vital-Durand, F. (1979). Development of the neural basis of visual acuity in monkeys: speculation on the origin of deprivation amblyopia. *Transactions of the Ophthalmological Society, U.K., 99*, 363-8.

Blakemore, C. & Vital-Durand, F. (1983). Development of contrast sensitivity by neurones in monkey striate cortex. *Journal of Physiology, 334*, 18-19P.

Blakemore, C. & Vital-Durand, F. (1986a). Effects of visual deprivation on the development of the monkey's lateral geniculate nucleus. *Journal of Physiology, 380*, 493-511.

Blakemore, C. & Vital-Durand, F. (1986b). Organization and postnatal development of the monkey's lateral geniculate nucleus. *Journal of Physiology, 380*, 453-91.

Blakemore, C. & Vital-Durand, F. (1992). Different neural origins for 'blur' amblyopia and strabismic amblyopia. *Ophthalmic Physiological Optics, 12, 83.*

Bloomfield, S.A., Xin, D. & Persky, S.E. (1995). A comparison of receptive field and tracer coupling size of horizontal cells in the rabbit retina. *Visual Neuroscience 12,* 985-999.

Bode-Greuel, K.M. & Singer W. (1988). Developmental changes of the distribution of binding sites for organic Ca2+-channel blockers in cat visual cortex. *Experimental Brain Research, 70,* 266-75.

Bode-Greuel, K.M. & Singer, W. (1989). The development of N-methyl-D-aspartate receptors in cat visual cortex. *Developmental Brain Research, 46,* 197-204.

Bolster B., Crowne D.P. (1979). Effects of anterior and posterior inferotemporal lesions on discrimination reversal in the monkey. *Neuropsychologia, 17,* 11-20.

Boothe, R.G., Dobson, V., & Teller, D.Y. (1985). Postnatal development of vision in human and nonhuman primates. *Annual Review of Neuroscience, 8,* 495-545.

Boothe, R.G., Greenough, W.T., Lund, J.S., & Wrege, K. (1979). A quantitative investigation of spine and dendrite development of neurons in visual cortex (Area 17) of macaca nemestrina monkeys. *Journal of Comparative Neurology, 186,*473-490.

Bourtchuladze, R., Frenguelli, B., Blendy, J., Cioffi, D., Schutz, G. & Silva, A. (1994). Deficient long-term memory in mice with a targeted mutation of the cAMP-responsive element-binding protein. *Cell, 79,* 59-68.

Boycott, B.B., Hopkins, J.M., & Sperling, H.G. (1987). Cone connections of the horizontal cells of the rhesus monkey's retina. *Proceedings of the Royal Society, London B, 229,* 345-79.

Braddick, O.J., Atkinson, J., Julesz, B., Kropfl, W., Bodis-Wollner, I., & Raab, E. (1980) Cortical binocularity in infants. *Nature, 288,* 363-365.

Brocher, S., Artola, A. & Singer, W. (1992). Agonists of cholinergic and noradrenergic receptors facilitate synergistically the induction of long-term potentiation in slices of rat visual cortex. *Brain Research, 573,* 27-36.

Burkhalter, A. & Bernardo, K.L. (1989) Organization of corticocortical connections inhuman visual cortex. *Proceedings of the National Academy of Science USA, 86,* 1071-1075.

Butler, T. & Westheimer, G. (1978). Interference with stereoscopic acuity: Spatial temporal and disparity tuning. *Vision Research, 18,* 1387-1397.

Cabelli, R.J., Allendoerfer, K.J., Radeke, M.J., Welcher, A.A., Feinstein, S.C. & Shatz, C.J. (1996). Changing patterns of expression and subcellular localization of TrkB in the developing visual system. *Journal of Neuroscience, 16,* 7965-80.

Calkins, D.J., Schein, S.J., Tsukamoto, Y., & Sterling, P. (1994). M and L cones in macaque fovea connect to midget ganglion cells by different numbers of excitatory synapses. *Nature, 371,* 70-72.

Callaway, E.M. & Katz, L.C. (1990). Emergence and refinement of clustered horizontal connections in cat striate cortex. *Journal of Neuroscience, 10,* 1134-53.

Carlson, M., Hubel, D.H., & Wiesel, T.N. (1986). Effects of monocular exposure to oriented lines on monkey striate cortex. *Developmental Brain Research, 25,* 71-81.

Catalano, S.M. & Shatz, C.J. (1998) Activity-dependent cortical target selection by thalamic axons. *Science, 281,* 559-562.

Chapman, B. & Godecke, I. (2000). Cortical cell orientation selectivity fails to develop in the absence of ON-center retinal ganglion cell activity. *Journal of Neuroscience, 20,* 1922-1930.

Chapman, B., Jacobson, M.D., Reiter, H.O., & Stryker, M.P. (1986). Ocular dominance shift in kitten visual cortex caused by imbalance in retinal electrical activity. *Nature, 324,* 154-6.

Chapman, B. & Stryker, M.P. (1993). Development of orientation selectivity in ferret visual cortex and effects of deprivation. *Journal of Neuroscience, 13,* 5251-62.

Cheng K., Hasegawa T., Saleem K.S., & Tanaka K. (1994). Comparison of neuronal selectivity for stimulus speed, length, and contrast in the prestriate visual cortical areas V4 and MT of the macaque monkey. *Journal of Neurophysiology, 71,* 2269-80.

Chung, S. & Ferster, D. (1998). Strength and orientation tuning of the thalamic input to simple cells revealed by electrically evoked cortical suppression. *Neuron, 20,* 1177-1189.

Cleland, B.G., Dubin, M.W., & Levick, W.R. (1971). Sustained and transient neurons in the cat's retina and lateral geniculate nucleus. *Journal of Physiology, 217,* 473-496.

Cohen-Cory, S. & Fraser, S.E. (1995). Effects of brain-derived neurotrophic factor on optic axon branching and remodeling in vivo. *Nature, 378,* 192-196.

Colby, C.L., Duhamel, J.R. & Goldberg, M.E. (1993) Ventral intraparietal area of the macaque: Anatomic location and visual response properties. *Journal of Neurophysiology, 69,* 902-914.

Colby, C.L., Duhamel, J.R. & Goldberg, M.E. (1995) Oculocentric spatial representations in parietal cortex. *Cerebral Cortex, 5,* 470-481.

Constantine-Paton, M., Cline, H.T., & Debski, E. (1990). Patterned activity, synaptic convergence, and the NMDA receptor in developing visual pathways. *Annual Review of Neuroscience, 13,* 129-154.

Courage, M.L. & Adams, R.J. (1990). Visual acuity assessment from birth to three years using the acuity card procedure: Cross sectional and longitudinal samples. *Optometry and Vision Science, 67,* 713-718.

Crair, M.A.C., Glimpse, D.C. & Stryker, M.P. (1998) The role of visual experience in the development of columns in cat visual cortex. *Science, 279,* 566-570.

Cramer, K.S, Angelic, A., Ham, J-O., Bogdanov, M.B. & Sur, M. (1996). A role for nitric oxide in the development of the ferret retinogeniculate projection. *Journal of Neuroscience, 16,* 7995-8004.

Crawford, M.L.J. & von Noorden, G.K. (1980). Optically induced concomitant strabismus in monkeys. *Investigative Ophthalmology & Visual Science, 19,* 1105-1109.

Crawford, M.L.J., Smith III, E.L., Harwerth, S., & von Noorden, G.K. (1984). Stereoblind monkeys have few binocular neurons. *Investigative Ophthalmology and Visual Science, 25,* 779-781.

Crawford, M.L.J., von Noorden, G.K., Meharg, L.S., Rhodes, J.W., Harwerth, R.S., Smith, E.L., & Miller, D.D. (1983). Binocular neurons and binocular function in monkeys and children. *Investigative Ophthalmology & Visual Science, 24,* 491-495.

Creutzfeldt, O.D., Kuhnt, U., & Benevento, L.A. (1974). An intracellular analysis of visual cortical neurons to moving stimuli: Responses in a cooperative neuronal network. *Experimental Brain Research, 21,* 251-274.

Crick, F.H.C. & Watson, J.D. (1954). The complementary structure of Deoxyribonucleic Acid. *Proceedings of the Royal Society of London, 223*, 80-96.

Crook, J.M., Lange-Malecki, B., Lee, B.B., & Valberg, A. (1988). Visual resolution of macaque retinal ganglion cells. *Journal of Physiology, 396*, 205-24.

Crowly, J.C. & Katz, L.C. (1999). Development of ocular dominance columns in the absence of retinal input. *Nature Neuroscience, 2*, 1125-30.

Cynader, M. & Regan, D. (1978). Neurones in cat prestriate cortex sensitive to the direction of motion in three-dimensional space. *Journal of Physiology, 274*, 549-569.

Dacey, D.M. (1996). Circuitry for color coding in the primate retina. *Proceedings of the National Academy of Science (USA), 93*, 582-588.

Dacey, D.M (1999). Primate retina: cell types, circuits, and color opponency. *Progress in Retinal and Eye Research, 18,* 737-763

Dacey, D.M. (2000). Parallel pathways for spectral coding in primate retina. *Annual Review of Neuroscience, 23*, 743-775.

Dacey, D.M., Lee, B.B., Stafford, D.K., Pokorny, J. & Smith, V.C. (1996). Horizontal cells of the primate retina: cone specificity without spectral opponency. *Science, 271*, 656-659.

Damasio, A.R. (1990). Category-related recognition defects as a clue to the neural substrates of knowledge. *Trends in Neurosciences, 13*, 95-98.

Damasio, A.R., Damasio, H., Van Hoesen, G.W. (1982). Prosopagnosia: Anatomic basis and behavioral mechanisms. *Neurology, 32*, 331-341.

Damasio, A.R., Tranel, D., & Damasio, H. (1990). Face agnosia and the neural substrates of memory. *Annual Review of Neuroscience, 13*, 89-109.

Damasio, A.R., Yamada, T., Damasio, H., Corbet, J., & McKee, J. (1980). Central achromatopsia: Behavioral anatomic and physiological aspects. *Neurology, 30*, 1064-1071.

de Courten, C. & Garey L.J. (1982). Morphology of the neurons in the human lateral geniculate nucleus and their normal development. A Golgi study. *Experimental Brain Research, 47*, 159-71.

De Valois, R.L., Abramov, I., & Jacobs, G.H. (1966). Analysis of response patterns of LGN cells. *Journal of the Optometry Society of America, 56*, 966-977.

De Valois, R.L., Yund, E.W., & Kepler, N.K. (1982). The orientation and direction selectivity of cells in macaque visual cortex. *Vision Research, 22*, 531-544.

Decaban, A. (1954). Human thalamus. An anatomical, developmental, and pathological study. *Journal of Comparative Neurology, 100*, 63.

Demb, J.B., Boynton, G. M., & Heeger, D.J. (1998). Functional magnetic resonance imaging of early visual pathways in dyslexia. *Journal of Neuroscience, 18*, 6939-6951.

Desimone, R., Schein, S.J., Moran, J., & Ungerleider, L.G. (1985). Contour, color, and shape analysis beyond the striate cortex. *Vision Research, 25*, 441-452.

Desmond, N.L. & Levy, W.B. (1986). Changes in the postsynaptic density with long-term potentiation in the dentate gyrus. *Journal of Comparative Neurology, 253*, 476-82.

DeYoe, E.A., & Van Essen, D.C. (1985). Segregation of efferent connections and receptive field properties in visual area V2 of the macaque. *Nature, 317*, 58-61.

Dreher, B., Fukada, Y., & Rodieck, R. W. (1976). Identification, classification, and anatomical segregation of cells with X-like and Y-like properties in the lateral geniculate nucleus of Old-World primates. *Journal of Physiology, 258*, 433-452.

Dreifuss, K., Kelly, J.S., & Krnjevic, (1969). Cortical inhibition and γ-aminobutyric acid. *Experimental Brain Research, 9*, 137-154.

Dubin, M.W., Stark, L.A., & Archer, S.M. (1986). A role for action-potential activity in the development of neuronal connections in the kitten retinogeniculate pathway. *Journal of Neuroscience, 6*, 1021-36.

Dudek, S.M. & Friedlander, M.J. (1996). Developmental down-regulation of LTD in cortical layer IV and its independence of modulation by inhibition. *Neuron, 16*, 1-20.

Duffy, C.J. & Wurtz, R.H. (1991a). Sensitivity of MST neurons to optic flow stimuli. I. A continuum of response selectivity to large-field stimuli. *Journal of Neurophysiology, 65*, 1329-45.

Duffy, C.J. & Wurtz, R.H. (1991b). Sensitivity of MST neurons to optic flow stimuli. II. Mechanisms of response selectivity revealed by small-field stimuli. *Journal of Neurophysiology, 65*, 1346-59.

Duhamel, J.R., Bremmer, F., BenHamed, S., & Graf, W. (1997) Spatial invariance of visual receptive fields in parietal cortex neurons. *Nature, 389*, 845-848.

Dursteler, M.R., & Wurtz, R.H. (1988). Pursuit and optokinetic deficits following chemical lesions of cortical areas MT and MST. *Journal of Neurophysiology, 60*, 940-65.

Ellis, A.W., Young, A.W. (1988). *Human Cognitive Neuropsychology.* Hove: Lawrence Erlbaum.

Enroth-Cugell, C. & Robson, J.G. (1966). The contrast sensitivity of retinal ganglion cells of the cat. *Journal of Physiology, 187*, 517-552.

Errington, M.L., Lynch, M.A., & Bliss, T.V.P. (1987). Long-term potentiation in the dentate gyrus: induction and increased glutamate release are blocked by D-aminophosphonovalerate. *Neuroscience, 20*, 279-284.

Eskandar, E.N. & Assad, J.A. (1999) Dissociation of visual, motor and predictive signals in parietal cortex during visual guidance. *Nature Neuroscience, 2*, 88-93.

Euler, T. & Masland, R.H. (2000). Light-evoked responses of bipolar cells in a mammalian retina. *Journal of Neurophysiology, 83*, 1817-1829.

Euler, T., Schneider, H. & Wässle, H. (1996). Glutamate responses of bipolar cells in a slice preparation of the rat retina. *Journal of Neuroscience, 16*, 2934-2944.

Felleman, D.J., Burkhalter, A. & Van Essen, D.C. (1997) Cortical connections of areasV3 and VP of macaque monkey extrastriate visual cortex. *Journal of Comparative Neurology, 379*, 21-47.

Felleman, D.J. & Van Essen, C. (1987). Receptive field properties of neurons in area V3 of macaque monkey extrastriate cortex. *Journal of Neurophysiology, 57*, 889-920.

Felleman, D.J. & Van Essen, D.C. (1991). Distributed hierarchical processing in the primate cerebral cortex. *Cerebral Cortex, 1*, 1-47.

Felleman, D.J., Xiao, Y. & McClendon, E. (1997). Modular organization of occipito-temporal pathways: cortical connections between visual area 4 and visual area 2 and posterior inferotemporal ventral area in macaque monkeys. *Journal of Neuroscience, 17*, 3185-3200.

Ferrera, V.P., Nealey, T.A., & Maunsell, J.H.R. (1991). Magnocellular and parvocellular inputs to macaque area V4. *Investigative Ophthalmology and Visual Science, 32,* 1117.

Ferrera, V.P., Nealey, T.A., Maunsell, J.H.R. (1994). Responses in macaque visual area V4 following inactivation of the parvocellular and magnocellular LGN pathways. *Journal of Neuroscience, 14,* 2080-8.

Ferster, D. (1986). Orientation selectivity of synaptic potentials in neurons of cat primary visual cortex. *Journal of Neuroscience, 6,* 1284-301.

Ferster, D. & Lindstrom, S. (1983). An intracellular analysis of geniculocortical connectivity in area 17 of the cat. *Journal of Physiology, 342,* 181-215.

Ferster, D. & Miller, K.D. Neural mechanisms of orientation selectivity in the visual cortex. *Annual Review of Neuroscience, 23,* 441-471; 2000.

Fiorentini, A., Baumgartner, G., Magnussen, S., Schiller, P. H., & Thomas, J. P. (1990). The perception of brightness and darkness: Relations to neuronal receptive fields. *Visual Perception: The Neurophysiological Foundations.* In Academic Press, pp.129-161.

Fitzpatrick, D., Lund, J.S., & Blasdel, G.G. (1985). Intrinsic connections of macaque striate cortex. Afferent and efferent connections of lamina 4C. *Journal of Neuroscience, 5,* 3329-3349.

Fox, K., Sato, H., & Daw, N. (1989). The location and function of NMDA receptors in cat and kitten visual cortex. *Journal of Neuroscience, 9,* 2443-54.

Fox, R., Aslin, R.N., Shea, S.L., & Dumais, S.T. (1979). Stereopsis in human infants. *Science, 207,* 323-324.

Fregnac, Y. & Imbert, M. (1984). Development of neuronal selectivity in primary visual cortex of cat. *Physiological Review, 64,* 325-434.

Fregnac, Y., Shulz, D., Thorpe, S., & Bienenstock, E. (1988). A cellular analogue of visual cortical plasticity. *Nature, 333,* 367-70.

Fregnac, Y., Shulz, D., Thorpe, S. & Bienenstock, E. (1992). Cellular analogs of visual cortical epigenesis: I. Plasticity of orientation selectivity. *Journal of Neuroscience, 12,* 1280-1300.

Freund, T.F., Martin, K.A.C., Smith, A.D., & Somogyi, P. (1983). Glutamate decarboxylase-immunoreactive terminals of Golgi-impregnated axo-axonic cells and of presumed basket cells of the cat's visual cortex. *Journal of Comparative Neurology, 221,* 263-278.

Fujita, I., Tanaka, K., Ito, M., & Ceng, K. (1992). Columns for visual features of objects in monkey inferotemporal cortex. *Nature, 360,* 343-346.

Galuske, R.A.W. & Singer, W. (1996). The origin and topography of long-range intrinsic projections in cat visual cortex: a developmental study. *Cerebral Cortex, 6,* 417-430.

Garey, L. J. & de Courten, C. (1983). Structural development of the lateral geniculate nucleus and visual cortex in monkey and man. *Behavioral Brain Research, 10,* 3.

Garey, L.J. & Saini, K.D. (1981). Golgi studies of the normal development of neurons in the lateral geniculate nucleus of the monkey. *Experimental Brain Research, 44,* 117-28.

Gawne, T. J. & Richmond, B. J. (1993). How independent are the messages carried by adjacent inferior temporal cortical neurons? *Journal of Neuroscience, 13,* 2758-2771.

Georgopoulos, A.P., Schwartz, A.B., & Kettner, R.E. (1986). Neuronal population coding of movement direction. *Science, 233,* 1416-9.

Geschwind, N. (1965). Disconnection syndromes in animals and man. *Brain, 88,* 237-294 and 585-644.

Geyer, O., Malach, R., & Sagi, D. (1991). Direct demonstration of the relationship between cytochrome oxidase (CO) dense blobs and dendritic arbors in monkey striate cortex. *Investigative Ophthalmology and Visual Science, 32,* 1118.

Ghose, G.M. & Ts'O, D. Y. (1997). Form processing modules in primate area V4. *Journal of Neurophysiology, 77,* 2191-2196.

Gilbert, C. D. & Wiesel, T. N. (1989). Columnar specificity of intrinsic horizontal and corticocortical connections in the cat visual cortex. *Journal of Neuroscience, 9,* 2432-2442.

Gilbert, C. D. & Wiesel, T. N. (1994). Circuitry and functional dynamics of adult visual cortex. In B. Albowitz, K. Albus, U. Kuhnt, H. -Ch. Nothdurft, & P. Wahle (Eds.), *Structural and Functional Organization of the Neocortex.* Berlin: Springer Verlag, pp. 227-240.

Girard, P. & Bullier, J. (1989). Visual activity in area V2 during reversible inactivation of area 17 in the macaque monkey. *Journal of Neurophysiology, 62,* 1287-1302.

Girard, P., Salin, P. A., & Bullier, J. (1991). Visual activity in macaque area V4 depends on area 17 input. *NeuroReport, 2,* 81-84.

Gnadt, J.W. & Andersen, R.A. (1988). Memory related motor planning activity in posterior parietal cortex of macaque. *Experimental Brain Research, 70,* 216-220.

Godecke, I. & Bonhoeffer, T. (1996). Development of identical orientation maps for two eyes without common visual experience. *Nature, 379,* 251-254.

Gottlieb, M.D., Pasik, P., & Pasik, T. (1985). Early postnatal development of the monkey visual system. I. Growth of the lateral geniculate nucleus and striate cortex. *Brain Research, 349,* 53-62.

Greenough, W.T., Hwang, H. -M., & Gorman, C. (1985). Evidence for active synapse formation, or altered postsynaptic metabolism, in visual cortex of rats reared in complex environments. *Proceedings of the National Academy of Sciences, 82,* 4549-4552.

Grunert, U. & Wassle, H. (1996). Glycine receptors in the rod pathway of the macaque monkey retina. *Visual Neuroscience, 13,* 101-115.

Gu, Q.A., Bear, M.F., Singer, W. (1989). Blockade of NMDA-receptors prevents ocularity changes in kitten visual cortex after reversed monocular deprivation. *Developmental Brain Research, 47,* 281-8.

Guillery, R. W. (1988). Competition in the development of the visual pathways. In J. G. Parnavelas, C. D. Stern, & R. V. Stirling (Eds.), *The making of the nervous system.* Oxford: Oxford University Press, pp.356-379.

Gustafsson, B., Wingstrom, H., Abraham, W. C., & Huang, Y. Y. (1987). Long-term potentiation in the hippocampus using depolarizing current pulses as the conditioning stimulus to single volley synaptic potentials. *Journal of Neuroscience, 7,* 774-780.

Hammond, P. Visual cortical processing: Textural sensitivity and its implications for classical views. In D. Rose & V. G. Dobson (Eds.), *Models of Visual Cortex.* Chichester: John Wiley, pp. 326-333.

Hammond, P. & MacKay, D. M. (1975). Differential responses of cat visual cortical cells to textured stimuli. *Experimental Visual Research, 22,* 427-430.

Harris, W.A. (1984). Axonal pathfinding in the absence of normal pathways and impulse activity. *Journal of Neuroscience, 4,* 1153-62.

Hata, Y., Tsumoto, T., Sato, H, & Tamura, H. (1991). Horizontal interactions between visual cortical neurons studied by cross-correlation analysis in the cat. *Journal of Physiology, 441,* 593-614.

Hawken, M.J., Parker, A.J., & Lund, J.S. (1988). Laminar organization and contrast sensitivity of direction-selective cells in the striate cortex of the Old-World monkey. *Journal of Neuroscience, 8,* 3541-3548.

Hebb, D. O. (1949). *The organization of behaviour.* New York: Wiley.

Held, R., Birch, E.E., & Gwiazda, J. (1980). Stereoacuity of human infants. *Proceedings of the National Academy of Sciences (USA), 77,* 5572-5574.

Hendrickson, A. & Rakic, P. (1977). Histogenesis and synaptogenesis in the dorsal lateral geniculate nucleus (LGN) of the fetal monkey brain. *Anatomical Records, 187,* 602.

Hendry, S.H. & Calkins, D.J. (1998). Neuronal chemistry and functional organization in the primate visual system. *Trends in Neuroscience, 21,* 344-9.

Hendry, S.H. & Reid, R. C. (2000). The koniocellular pathway in primate vision. *Annual Review of Neuroscience, 23,* 127-153.

Hendry, S.H. & Yoshioka, T. (1994). A neurochemically distinct third channel in the macaque dorsal lateral geniculate nucleus. *Science, 264,* 575-7.

Heywood, C.A. & Cowey, A. (1987). On the role of cortical area V4 in the discrimination of hue and pattern in macaque monkeys. *Journal of Neuroscience, 7,* 2601-17

Heywood, C. A. & Cowey, A. (1992). The role of the "face-cell" area in the discrimination and recognition of faces in monkeys. *Philosophical Transactions of the Royal Society, London, B., 335,* 31-37.

Heywood, C. A. & Cowey, A. (1993). Colour and face perception in man and the monkey: The missing link. In B. Gulyas, D. Ottoson, & P. E. Roland (Eds.), *Functional organization of the human visual cortex.* Oxford: Pergamon Press, pp. 195-210. .

Hickey, T.L. (1977). Postnatal development of the human lateral geniculate nucleus: Relationship to a critical period for the visual system. *Science, 198,* 836.

Hickey, T.L. & Peduzzi, J.D. (1987). Structure and development of the visual-system. *Handbook of infant development,* vol. 1. Academic Press, pp. 1-42.

Hirsch, H.V.B. & Spinelli, D.N. (1970). Visual experience modifies distribution of horizontally and vertically oriented receptive fields in cats. *Science, 168,* 869-71.

Hirsch, J. A., Alonso, J. M., Reid, R. C., & Martinez, L. M. (1998). Synaptic integration in striate cortical simple cells. *Journal of Neuroscience, 18,* 9517-9528.

Hitchcock, P.F. & Hickey, T. L. (1980). Prenatal development of the human lateral geniculate nucleus. *Journal of Comparative Neurology, 194,* 395.

Horton, J.C. & Hocking, D.R. (1997). Timing of the critical period for plasticity of ocular dominance columns in macaque striate cortex. *Journal of Neuroscience, 17,* 3684-709.

Horton, J.C. & Hocking, D.R. (1998). Monocular core zones and binocular border strips in primate striate cortex revealed by the contrasting effects of enucleation, eyelid suture, and retinal laser lesions on cytochrome oxidase activity. *Journal of Neuroscience, 18,* 5433-55.

Hubel, D.H. & Livingstone, M.S. (1985). Complex-unoriented cells in a subregion of primate area 18. *Nature, 315,* 325-327

Hubel, D.H. & Livingstone, M.S. (1987). Segregation of form, color, and stereopsis in primate area 18. *Journal of Neuroscience, 7,* 3378-3415.

Hubel, D.H. & Wiesel, T.N. (1962). Receptive fields, binocular interaction, and functional architecture in the cat's visual cortex. *Journal of Physiology, 160,* 106-154.

Hubel, D.H. & Wiesel, T.N. (1963). Receptive fields of cells in striate cortex of very young, visually inexperienced kittens. *Journal of Neurophysiology, 26,* 994-1002.

Hubel, D.H. & Wiesel, T.N. (1968). Receptive fields and functional architecture of monkey striate cortex. *Journal of Physiology, 195,* 215-243.

Hubel, D.H. & Wiesel, T.N. (1970). The period of susceptibility to the physiological effects of unilateral eye closure in kittens. *Journal of Physiology, 206,* 419-436.

Hubel, D.H. & Wiesel, T.N. (1972). Laminar and columnar distribution of geniculocortical fibres in the macaque monkey. *Journal of Comparative Neurology, 146,* 421-450.

Huttenlocher, P.R. (1990). Morphometric study of human cerebral cortex development. *Neuropsychologia, 28,* 517-527.

Huttenlocher, P.R., de Courten, C., Garey, L.G., & Van der Loos, H. (1982). Synaptogenesis in human visual cortex -- evidence for synapse elimination during normal development. *Neuroscience Letters, 33,* 247-52.

Jacobson, S.G., Mohindra, I., & Held, R. (1981) Development of visual acuity in infants with congenital cataracts. *British Journal of Ophthalmology, 65,* 727-35.

Jacoby, R.A., Wiechmann, A.F., Amara, S.G., Leighton, B.H. & Marshak, D.W. (2000). Diffuse bipolar cells provide input to off parasol ganglion cells in the macaque retina. *Journal of Comparative Neurobiology, 46,* 6-18.

Jeannerod, M., Arbib, M.A., Rizzolatti, G., & Sakata, H. (1995). Grasping objects: the cortical mechanisms of visuomotor transformation. *Trends in Neurosciences, 18,* 314-20.

Julesz, B. (1978). Global stereopsis: Cooperative phenomena in stereoscopic depth perception. In R. Held, H.W. Leibowitz, & H.L. Teuber, (Eds.), *Handbook of Sensory Physiology VIII. Perception*, pp. 215-256. New York: Springer-Verlag.

Julesz, B. (1986). Stereoscopic vision. *Vision Research, 26,* 1601-1612.

Kaas, J. (1993). The organization of visual cortex in primates: Problems, conclusions, and the use of comparative studies in understanding the human brain. In B. Gulyas, D. Ottolson, & P. E. Roland (Eds.). *Functional organization of the human visual cortex.* Oxford: Pergamon Press, pp. 1-11.

Kalil, R.E., Spear, P.D., & Langtsetmo, A. (1984). Response properties of striate cortex neurons in cats raised with convergent strabismus. *Journal of Neurophysiology, 52,* 514-537.

Kamermans, M. & Spekreijse, H. (1999) The feedback pathway from horizontal cells to cones: A mini review with a look ahead. *Vision Research, 39,* 2449-2468.

Kandel, E.R. (1991). Perception of motion, depth, and form. In E.R. Kandel, J.H. Schwartz, & T.M. Jessell (Eds.), *Principles of neural science, 3rd Edition*. New York: Elsevier, pp. 441-466.

Kasamatsu, T. & Petigrew, J.D. (1976). Depletion of brain catecholamines: Failure of ocular dominance shift after monocular occlusion in kittens. *Science, 194*, 206-9.

Kasamatsu, T., Petigrew, J.D., & Ary, M. (1979). Restoration of visual cortical plasticity by local microperfusion of norepinephrine. *Journal of Comparative Neurology*, 163-81.

Katz (1991). Specificity in the development of vertical connections in cat striate cortex. *European Journal of Neuroscience, 3*, 1-9.

Katz, L.C. & Callaway, E.M. (1992). Development of local circuits in mammalian visual cortex. *Annual Review of Neuroscience, 15*, 31-56.

Katz, L.C., Gilbert, C.D., & Wiesel, T.N. (1989). Local circuits and ocular dominance columns in monkey striate cortex. *Journal of Neuroscience, 9*, 1389-99.

Kelso, S. R., Ganong, A. H., & Brown, T. H. (1986). Hebbian synapses in hippocampus. *Proceedings of the National Academy of Sciences, USA, 83*, 5326-5330.

Kleinschmidt, A., Bear, M.F., & Singer, W. (1987). Blockade of "NMDA" receptors disrupts experience-dependent modifications of kitten striate cortex. *Science, 238*, 355-358.

Komatsu, H. & Ideura, Y. (1993). Relationships between color, shape, and pattern selectivities of neurons in the inferior temporal cortex of the monkey. *Journal of Neurophysiology, 70*, 677-694.

Koutalos, Y. & Yau, K.W. (1996). Regulation of sensitivity in vertebrate rod photoreceptors by calcium. *Trends in Neurosciences, 19*, 73-81.

Kratz, K.E., Spear, P.D., & Smith, D.C. (1976). Postcritical-period reversal of monocular deprivation on striate cortex cells in the cat. *Journal of Neurophysiology, 39*, 501-11.

LaChica, E.A., & Casagrande, V.A. (1988). Development of primate retinogeniculate axon arbors. *Visual Neuroscience, 1*, 103-23.

Lagae, L., Maes, H., Raiguel, S., Xiao, D.K., & Orban, G.A. (1994) Responses of macaque STS neurons to optic flow components: A comparison of area MT and MST. *Journal of Neurophysiology, 71*, 1597-1626.

Land, E.H. (1974). The retinex theory of colour vision. *Proceedings of the Royal Institute of Great Britain, 47*, 23-57.

La Vail, M.M., Rapaport, D.H., & Rakic, P. (1992). Cytogenesis in the monkey retina. *Journal of Comparative Neurology*, 322, 577-588.

Lee, B.B. (1996). Receptive field structure in the primate retina. *Visual Research, 36*, 631-44.

Lee, B.B., Martin, P.R., & Valberg, A. (1989). Sensitivity of macaque ganglion cells to luminance and chromatic flicker. *Journal of Physiology, 414*, 223-243.

Lein, E.S., Hohn, A. & Shatz, C.J. (2000). Dynamic regulation of BDNF and NT-3 expression during visual system development. *Journal of Comparative Neurology, 420*, 1-18.

Lein, E.S. & Shatz, C.J. (2000). Rapid regulation of brain-derived neurotrophic factor mRNA within eye-specific circuits during ocular dominance column formation. *Journal of Neuroscience, 20*, 1470-1483.

Leopold, D.A., Logothetis, N.K. (1996). Activity changes in early visual cortex reflect monkeys' percepts during binocular rivalry. *Nature, 379,* 549-53.

LeVay, S., Stryker, M.P., & Shatz, C.J. (1978). Ocular dominance columns and their development in layer IV of the cat's visual cortex: a quantitative study. *Journal of Comparative Neurology, 179,* 223-44.

LeVay, S., Wiesel, T.N, & Hubel, D.H. (1980). The development of ocular dominance columns in normal and visually deprived monkeys. *The Journal of Comparative Neurology, 191,* 1-51.

Leventhal, A.G. & Hirsch, H.V. (1977). Effects of early experience upon orientation sensitivity and binocularity of neurons in visual cortex of cats. *Proceedings of the National Academy of Sciences, USA, 74,* 1272-6.

Leventhal, A.G. & Hirsch, H.V. (1980). Receptive-field properties of different classes of neurons in visual cortex of normal and dark-reared cats. *Journal of Neurophysiology, 43,* 1111-32.

Levick, W. R. (1975). Form and function of cat retinal ganglion cells. *Nature, 254,* 659-662.

Levitt, J.B., Kiper, D.C., Movshon, J.A. (1994). Receptive fields and functional architecture of macaque V2. *Journal of Neurophysiology, 71,* 2517-42.

Lhermitte, F. & Beauvois, M.F. (1973). A visual-speech disconnexion syndrome. Report of a case with optic aphasia, agnosic alexia and colour agnosia. Brain, 96, 695-714.

Livingstone, M.S. & Hubel, D.H. (1984). Anatomy and physiology of a color system in the primate visual cortex. *Journal of Neuroscience, 4,* 309-356.

Livingstone, M.S. & Hubel, D.H. (1987). Connections between layer 4B of area 17 and the thick cytochrome oxidase stripes of area in the squirrel monkey. *Journal of Neuroscience, 7,* 3371-3377.

Logothetis, N.K. (1998). Single units and conscious vision. *Philosophical Transactions of the Royal Society London B., 353,* 1801-18.

Logothetis, N.K., Schiller, P.H., Charles, E.R. & Hurlbert, A.C. (1990). Perceptual deficits and the activity of the color-opponent and broad-band pathways at isoluminance. *Science, 247,* 214-217.

Lu, B., Yokoyama, M., Dreyfus, C.F. & Black, I.B. (1991). Depolarizing stimuli regulate nerve growth factor gene expression in cultured hippocampal neurons. *Proceedings of the National Academy of Sciences, USA, 88,* 6289-92.

Luhmann, H.J., Singer, W. & Martinez-Millan, L. (1991). Horizontal interactions in cat striate cortex: I. Anatomical substrate and postnatal development. *European Journal of Neuroscience, 2,* 344-357.

Lund, J.S. (1988). Anatomical organization of macaque monkey striate visual cortex. *Annual Review of Neuroscience, 11,* 253-88.

Lund, J.S., Boothe, R.G., & Lund, R.D. (1977). Development of neurons in the visual cortex of the monkey (Macaca nemestrina): a Golgi study from fetal day 127 to postnatal maturity. *Journal of Comparative Neurology, 176,* 149-88.

Lund, J.S. & Holbach, S.M. (1991). Postnatal development of thalamic recipient neurons in the monkey striate cortex: I. Comparison of spine acquisition and dendritic growth of layer 4C alpha and beta spiny stellate neurons. *Journal of Comparative Neurology, 309,* 115-128.

Maffei, L., Berardi, N., Domenici, L., Parisi, V. & Pizzorusso, T. (1992). Nerve growth factor (NGF) prevents the shift in ocular dominance distribution of visual cortical neurons in monocularly deprived fats. *Journal of Neuroscience, 12,* 4651-4662.

Magoon, E.H. & Robb, R.M. (1981). Development of myelin in human optic nerve and tract. *Archives of Ophthalmology, 99,* 655-.

Malach, R., Amir, Y., Harel, M., Grinvald, A. (1993). Relationship between intrinsic connections and functional architecture revealed by optical imaging and in vivo targeted biocytin injections in primate striate cortex. *Proceedings of the National Academy of Science, USA, 90,* 10469-73

Marr, D. (1982). *Vision: A computational investigation into the human representation and processing of visual information.* San Francisco: Freeman.

Martin, P.R. (1998). Color Processing in the primate retina: recent progress. *Journal of Physiology,* 513.3, 631-638.

Mason, C.A. & Sretavan, D.W. (1997). Glia, neurons, and axon pathfinding during optic chiasm development. *Current Opinions in Neurobiology, 7,* 647-653.

Mastrogrande, D.N. (1983). Correlated firing of cat retinal ganglion cells. I. Spontaneously active inputs to X- and Y-cells. *Journal of Neurophysiology, 49,* 303-324.

Maunsell, J.H.R., Nealey, T.A., & DePriest, D.D. (1990). Magnocellular and parvocellular contributions to responses in the Middle Temporal visual area (MT) of the macaque monkey. *Journal of Neuroscience, 10,* 3323-3334.

Maunsell, J.H.R. & Newsome, W.T. (1987). Visual processing in monkey extrastriate cortex. *Annual Review of Neuroscience, 10,* 363-401.

McAllister, A.K., Katz, L.C. & Lo, D.C. (1996). Neurotrophin regulation of cortical dendritic growth requires activity. *Neuron, 17,* 1057-1064.

McAllister, A.K., Katz, L.C. & Lo, D.C. (1997). Opposing roles for endogenous BDNF and NT-3 in regulating cortical dendritic growth. *Neuron, 18,* 767-78.

McAllister, A.K., Lo, D.C. & Katz, L.C. (1995). Neurotrophins regulate dendritic growth in developing visual cortex. *Neuron, 15,* 791-803.

McCarthy R.A. & Warrington, E.K. (1986). Visual associative agnosia: a clinico-anatomical study of a single case. *Journal of Neurology, Neurosurgery, and Psychiatry 49,* 1233-40.

McClelland, J.L. & Rumelhart, D.E. (1981). An interactive activation model of context effects in letter perception: Part I. An account of basic findings. *Psychological Review, 88,* 375-405.

McGuire, B.A., Hornung, J.P., Gilbert, C.D., & Wiesel, T.N. (1984). Patterns of synaptic input to layer 4 of cat striate cortex. *Journal of Neuroscience, 4,* 3021-3033.

McKeefry, D. J. & Zeki, S. (1997). The position and topography of the human colour center as revealed by functional magnetic resonance imaging. *Brain, 120,* 2229-2242.

Meister, M. (1996). Multineuronal codes in retinal signaling. *Proceedings of the National Academy of Sciences, USA, 93,* 609-614.

Meister, M., Wong, R.O.L., Baylor, D.A., & Shatz, C.J. (1991). Synchronous bursts of action potentials in ganglion cells of the developing mammalian retina. *Science, 252,* 939-943.

Merigan, W.H. (1989). Chromatic and achromatic vision of macaques: role of the P pathway. *Journal of Neuroscience, 9,* 776-83.

Merigan, W.H., Byrne, C.E., Maunsell, J.H.R. (1991). Does primate motion perception depend on the magnocellular pathway? *Journal of Neuroscience, 11,* 3422-9.

Merigan, W.H. & Maunsell, J.H.R. (1993). How parallel are the primate visual pathways? *Annual Review of Neuroscience, 16,* 369-402.

Michael, C.R. (1985). Laminar segregation of color cells in the monkey's striate cortex. *Vision Research, 25,* 415-423.

Michael, C.R. (1988). Retinal afferent arborization patterns, dendritic field orientations, and the segregation of function in the lateral geniculate nucleus of the monkey. *Proceedings of the National Academy of Sciences, USA, 85,* 4914-4918.

Mikami, A., Newsome, W.T., & Wurtz, R.H. (1986a). Motion selectivity in macaque visual cortex. II. Spatiotemporal range of directional interactions in MT and V1. *Journal of Neurophysiology, 55,* 1328-1339.

Mikami, A., Newsome, W.T., & Wurtz, R.H. (1986b). Motion selectivity in macaque visual cortex. I. Mechanisms of direction and speed selectivity in extrastriate area MT. *Journal of Neurophysiology, 55,* 1308-1327.

Mitchel, A.E. & Garey, L.J. (1984). The development of dendritic spines in the human visual cortex. *Human Neurobiology, 3,* 223-227.

Mitchel, D. E., Giffen, F., & Timney, B. (1976). A behavioral technique for the rapid assessment of the visual capabilities of kittens. *Perception, 6,* 181-193.

Miyashita, Y. (1988). Neuronal correlate of visual associative long-term memory in the primate temporal cortex. *Nature, 335,* 817-820.

Miyashita, Y. (1993). Inferior temporal cortex: Where visual perception meets memory. *Annual Review of Neuroscience, 16,* 245-263.

Miyashita, Y. & Chang, H. S. (1988). Neuronal correlate of pictorial short-term memory in the primate temporal cortex. *Nature, 331,* 68-70.

Moran, J. & Desimone, R. (1985). Selective attention gates visual processing in the extrastriate cortex. *Science, 229,* 782-784.

Movshon, J.A., Adelson, E.H., Gizzi, M.S., & Newsome, W.T. (1985). The analysis of moving visual patterns. In C. Chagas, R. Gattass, & C. Gross (Eds.), *Pattern Recognition Mechanisms.* Rome: Vatican Press.

Movshon, J.A., Eggers, H.M., Gizzi, M.S., Hendrickson, A.E., Kiorpes, L., Boothe, R.G. (1987). Effects of early unilateral blur on the macaque's visual system. III. Physiological observations. *Journal of Neuroscience, 7,* 1340-51.

Mower, G. D., Berry, D., Burchfiel, J. L., & Duffy, F. H. (1981). Comparison of the effects of dark rearing and binocular suture on development and plasticity in cat visual cortex. *Brain Research, 220,* 255-267.

Muller, J.F., & Dacheux, R.F. (1997). Alpha ganglion cells of the rabbit retina lose antagonistic surround responses under dark adaptation. *Visual Neuroscience, 14,* 395-401.

Nakamura, H., Gattass, R., Desimone, R., & Ungerleider, L.G. (1993). The modular organization of projections from areas V1 and V2 to areas V4 and TEO in macaques. *Journal of Neuroscience, 13,* 3681-91.

Nathans, J., Sung, C-H., Weitz, C.J., Davenport, C.M., Merbs, S.L., & Wang, Y. (1992). Visual pigments and inherited variation in human vision. In D.P. Corey and S.D.

Roper (Eds.), Sensory Transduction. Society of General Physiologists Series, Vol. 47. The Rockefeller University Press, pp. 109-131.

Nealey, T.A. & Maunsell, J.H.R. (1991). Magnocellular contributions to the superficial layers of macaque striate cortex. *Investigative Ophthalmology and Visual Science, 32*, 1116.

Newsome, W.T., Mikami, A., & Wurtz, R.H. (1986). Motion selectivity in macaque visual cortex. III. Psychophysics and physiology of apparent motion. *Journal of Neurophysiology, 55*, 1340-1251.

Nikara, T., Bishop, P. O., & Petigrew. J D. (1968). Analysis of retinal correspondence by studying receptive fields of binocular single units in cat striate cortex. *Experimental Brain Research, 6*, 353-372.

Norden, J. J., Lettes, A., Costello, B., Lin, L.-H., Wouters, B., Bock, S., & Freeman, J. A. (1991). Possible role of GAP-43 in calcium regulation/neurotransmitter release. *Annals of the New York Academy of Science, 627*, 75-93.

O'Kursky, J. & Colonnier, M. (1982). Postnatal changes in the number of neurons and synapses in the visual cortex (area 17) of the macaque monkey: a stereological analysis in normal and monocularly deprived animals. *Journal of Comparative Neurology, 210*, 91-306.

Olson, C.R. & Freeman, R.D. Progressive changes in kitten striate cortex during monocular vision. *Journal of Neurophysiology, 38*, 26-32.

Pandya, D. & Yeterian, E. H. (1985). Architecture and connections of cortical association areas. In Peters & Jones (Eds.), *Cerebral Cortex, vol. 4*. New York: Plenum Press, pp. 3-64.

Papanicolaou, A. C. (1998). *Fundamentals of functional brain imaging: A guide to the methods and their applications to psychology and behavioral neurosciences*. Netherlands: Swets and Zeitlinger.

Perkins, T. A. & Teyler, T. J. (1988). A critical period for long-term potentiation in the developing rat visual cortex. *Brain Research, 439*, 222-229.

Perlman, I. & Normann, R.A. (1998). Light adaptation and sensitivity controlling mechanisms in vertebrate photoreceptors. *Progress in Retinal and Eye Research, 17*, 523-563.

Perrett, D.I., Harries, M., Bevan, R., Thomas, S., Benson, P.J., Mistlin, A.J., Chitty, P.A.J., Hietanen, J.K., & Ortega, J.E. (1989). Frameworks of analysis for the neural representation of animate objects and actions. *Journal of Experimental Biology, 146*, 87-113.

Perrett, D.I. & Mistlin, A.J. (1989). Perception of facial characteristics by monkeys. In M. Berkeley & W. Stebbins (Eds.), *Comparative Perception*. New York: John Wiley.

Perrett, D.I., Rolls, & Caan, W. (1982). Visual neurons responsive to faces in the monkey temporal cortex. *Experimental Brain Research, 47*, 329-342.

Perrett, D. I., Smith, P.A.J., Potter, D.D., Mistlin, A.J., Head, A.S., Milner, A.D., & Jeeves, M.A. (1984). Neurones responsive to faces in the temporal cortex: Studies of functional organization, sensitivity to identity and relation to perception. *Human Neurobiology, 3*, 197-208.

Perry, V. H., Oehler, R., & Cowey, A. (1984). Retinal ganglion cells that project to the dorsal lateral geniculate nucleus in the macaque monkey. *Neuroscience, 12*, 1101-1123.

Peters, A., Payne, B. R., & Budd, J. (1994). A numerical analysis of the geniculocortical input to striate cortex in the monkey. *Cerebral Cortex, 4,* 215-229.

Petrig, B., Julesz, B., Kropfl, W., Baumgartner, G., & Anliker, M. (1981). Development of stereopsis and cortical binocularity in human infants: Electrophysiological evidence. *Science, 213,* 1402-1405.

Pettigrew, J. D., Nikara, T., & Bishop, P. O. (1968). Binocular interaction on single units in cat striate cortex: Simultaneous stimulation by single moving slit with receptive fields in correspondence. *Experimental Brain Research, 6,* 391-410.

Pirchio, M., Spinelli, D., Fiorentini, A., & Maffei, L. (1978). Infant contrast sensitivity evaluated by evoked potentials. *Brain Research, 141,* 179-184.

Pizzorusso, T., Porciatti, V., Tseng, J.L., Aebischer, P. & Maffei, L. (1997). Transplant of polymer-encapsulated cells genetically engineered to release nerve growth factor allows a normal functional development of the visual cortex in dark-reared rats. *Neuroscience, 80,* 307-311.

Poggio, G.F., & Fischer, B. (1977). Binocular interaction and depth sensitivity of striate and prestriate cortical neurons of the behaving rhesus monkey. *Journal of Neurophysiology, 40,* 1392-1405.

Poggio, G.F., Motter, B.C., Squatrito, S., & Trotter, Y. (1985). Responses of neurons in visual cortex (V1 and V2) of the alert macaque to dynamic random-dot stereograms. *Vision Research, 25,* 397-406.

Provis, J.M., van Driel, D., Billson, F.A., Russell, P. (1985). Development of the human retina: patterns of cell distribution and redistribution in the ganglion cell layer. *Journal of Comparative Neurology, 233,* 429-51.

Ptito, A., Lepore, F., Ptito, M., & Lassonde, M. (1987). Target detection and movement discrimination in the blind field of hemispherectomized patients. *Brain, 114,* 497-512.

Purves, D., & Lamantia, A. (1993). Development of blobs in the visual cortex of macaques. *Journal of Comparative Neurology, 334,* 169-175.

Rakic, P. (1972). Mode of cell migration to the superficial layers of fetal monkey neocortex. *Journal of Comparative Biology, 145,* 61-84.

Rakic, P. (1976). Prenatal genesis of connections subserving ocular dominance in the rhesus monkey. *Nature, 261,* 467-71.

Rakic, P. (1977). Genesis of the dorsal lateral geniculate nucleus in the rhesus monkey: Site and time of origin, kinetics of proliferation, routes of migration and pattern of distribution of neurons. *Journal of Comparative Neurology, 176,* 23.

Rakic, P. (1991). Plasticity of cortical development. In Brauth, S.E., Hall, W.S., & Dooling, R.J. (Eds.), *Plasticity of development.* Cambridge, Mass: MIT Press, pp.127-162.

Rauschecker, J.P. (1991). Mechanisms of visual plasticity: Hebb synapses, NMDA receptors, and beyond. *Physiological Review, 71,* 587-615.

Reid, R.C. & Alonso, J.M. (1995). Specificity of monosynaptic connections from thalamus to visual cortex. *Nature, 378,* 281-284.

Reiter, H.O. & Stryker, M.P. (1988). Neural plasticity without postsynaptic action potentials: Less-active inputs become dominant when kitten visual cortical cells are pharmacologically inhibited. *Proceedings of the National Academy of Sciences, USA, 85,* 3623-7.

Richmond, B.J. & Sato, T. (1982). Visual responses of inferior temporal neurons are modified by attention to different stimulus dimensions. *Society of Neuroscience Abstracts, 8,* 812.

Richmond, B.J., Wurtz, R.H., & Sato, T. (1983). Visual responses of inferior temporal neurons in awake rhesus monkey. *Journal of Neurophysiology, 50,* 1415-1432.

Riddoch, M.J., Humphreys, G.W., Coltheart, M. & Funnell, E. (1988). Semantic systems or system? Neuropsychological evidence re-examined. *Cognitive Neuropsychology, 5,* 3-25.

Rizzo, M., Corbett, J.J., Thomson, H.S., & Damasio, A.R. (1986). Spatial contrast sensitivity in facial recognition. *Neurology, 36,* 1254-1256.

Rizzo, M., Nawrot, M., Blake, R., Damasio, A. (1992). A human visual disorder resembling area V4 dysfunction in the monkey. *Neurology, 42,* 1175-80.

Rocha-Miranda, C.E., Bender, D.B., Gross, C.G., & Miskin, M. (1975). Visual activation of neurons in the inferotemporal cortex depends on striate cortex and forebrain commissures. *Journal of Neurophysiology, 38,* 475-491.

Rodman, H. R., Gross, C.G., & Albright, T.D. (1989). Afferent basis of visual response properties in area MT of the macaque: II. Effects of superior colliculus removal. *Journal of Neuroscience, 10,* 2033-2050.

Roe, A.W. & Ts'o, D.Y. (1995). Visual topography in primate V2: multiple representation across functional stripes. *Journal of Neuroscience, 15,* 3689-715.

Roelfsema, P.R., Konig, P., Engel, A.K., Sireteanu, R., Singer, W. (1994). Reduced synchronization in the visual cortex of cats with strabismic amblyopia. *European Journal of Neuroscience, 6,* 1645-55.

Rolls, E.T. & Baylis, G.C. (1986). Size and contrast have only small effects on the responses to faces of neurons in the cortex of the superior temporal sulcus of the monkey. *Experimental Brain Research, 65,* 38-48.

Ruthazer, E.S. & Stryker, M.P. (1996). The role of activity in the development of long-range connections in area 17 of the ferret. *Journal of Neuroscience, 16,* 7253-7269.

Ruttiger, L., Braun, D. I., Gegenfurtner, K. R., Petersen, D., et al. (1999). Selective color constancy deficits after circumscribed unilateral brain lesions. *Journal of Neuroscience, 19,* 3094-3106.

Saito, H., Tanaka, K., Yasuda, M., & Mikami, A. (1989). Directionally selective response of cells in the middle temporal area (MT) of the macaque monkey to the movement of equiluminous opponent color stimuli. *Experimental Brain Research, 75,* 1-14.

Saito, H., Yukie, M., Tanaka, K., Hikosaka, K., Fukada, Y., & Iwai, E. (1986). Integration of direction signals of image motion in the superior temporal sulcus of the macaque monkey. *Journal of Neuroscience, 6,* 145-157.

Sakai, K. & Miyashita, Y. (1991). Neural organization for the long-term memory of paired associates. *Nature, 354,* 152-155.

Sakata, H., Taira, M., Kusunoki, M., Murata, A., Tanaka, Y. (1997). The TINS Lecture. The parietal association cortex in depth perception and visual control of hand action. *Trends in Neurosciences, 20,* 350-7.

Sakata, H., Taira, M., Kusunoki, M., Murata, A., Tanaka, Y. & Tsutsui, K. (1998) Neural coding of 3D features of objects for hand action in the parietal cortex of the monkey. *Philosophical Transactions of the Royal Society, London B,* 1363-1375.

Sala, R., Viegi, A., Rossi, F.M., Pizzorusso, T., Bonanno, G., Raiteri, M. & Maffei, L. (1998). NGF and BDNF increase transmitter release in the rat visual cortex. *European Journal of Neuroscience, 19,* 2185-2191.

Sartori, G. & Job, R. (1988). The oyster with four legs: A neuropsychological study on the interaction of visual and semantic information. *Cognitive Neuropsychology, 5,* 105-132.

Schein, S.F. & de Monasterio, F.M. (1987). Mapping of retinal and geniculate neurons onto striate cortex of macaque. *Journal of Neuroscience, 7,* 996-1009.

Schein, S.J. & Desimone, R. (1990). Spectral properties of V4 neurons in the macaque. *Journal of Neuroscience, 10,* 3369-3387.

Scherer, W.J. & Udin, S.B. (1991). Chronic effects of NMDA and APV on tectal output in Xenopus laevis. *Visual Neuroscience, 6,* 185-92.

Schiller, P.H. (1993). The effects of V4 and middle temporal (MT) area lesions on visual performance in the rhesus monkey. *Visual Neuroscience, 10,* 717-746.

Schiller, P.H. (1996). On the specificity of neurons and visual areas. *Behavioral Brain Research, 76,* 21-35.

Schiller, P.H. & Logothetis, N.K. (1990). The color-opponent and broad-band channels of the primate visual system. *Trends in Neurosciences, 13,* 392-398.

Schiller, P.H., Logothetis, N.K. & Charles, E.R. (1990) Functions of the colour-opponent and broad-band channels of the visual system. *Nature, 343,* 68-70.

Schiller, P.H., Logothetis, N. K. & Charles, E. R. (1990). Role of the color-opponent and broad –band channels in vision. *Visual Neuroscience, 5,* 321-46.

Schiller, P.H. & Logothetis, N.K. & Charles, E.R. (1991) Parallel pathways in the visual system: Their role in perception at isoluminance. *Neuropsychologia, 29,* 433-441.

Schmidt, J. T. (1983). Regeneration of the retinotectal projection following compression onto a half tectum in goldfish. *Journal of Embryology and Experimental Morphology, 7,* 39-51.

Schmidt, J. T. (1994). The roles of activity, competition, and continued growth in the formation and stabilization of retinotectal connections in fish and frog. In V. A. Casagrande (Ed.), *Advances in neural and behavioral development, vol. 4.* pp. 69-122.

Schmidt, J. T., Cicerone, C. M., & Easter, S. S. (1978). Expansion of the half retinal projection to the tectum in goldfish: An electrophysiological and anatomical study. *Journal of Comparative Neurology, 177,* 257-278.

Schmidt, J. T. & Eisele, L. E. (1985). Stroboscopic illumination and dark rearing block the sharpening of the retinotectal map in goldfish. *Neuroscience, 14,* 535-546.

Schmidt, J. T. & Eisele, L. E. (1988). Activity sharpens the regenerating retinotectal projection in goldfish: Sensitive period for strobe illumination and lack of effects on synaptogenesis and on ganglion cell receptive field properties. *Journal of Neurobiology, 19,* 395-411.

Schmidt, K.E., Galuske, R.A. & Singer, W. (1999). Matching the modules: cortical maps and long-range intrinsic connections in visual cortex during development. *Journal of Neurobiology, 41,* 10-17.

Schoups, A.A., Elliott, R.C., Friedman, W.J. & Black, I.B. (1995). NGF and BDNF are differentially modulated by visual experience in the developing geniculocortical pathway. *Developmental Brain Research, 86,* 326-334.

Sclar, G., Freeman, R.D. (1982). Orientation selectivity in the cat's striate cortex is invariant with stimulus contrast. *Experimental Brain Research, 46,* 457-61.

Seltzer, B. & Pandya, D. N. (1989). Intrinsic connections and architectonics of the superior temporal sulcus in the rhesus monkey. *Journal of Comparative Neurology, 290,* 451-471.

Sengpiel, F. & Blakemore, C. (1996). The neural basis of suppression and amblyopia in strabismus. *Eye, 1996,* 250-258.

Sengpiel, F., Blakemore, C., Kind, P.C., & Harrad, R. (1994). Interocular suppression in the visual cortex of strabismic cats. *Journal of Neuroscience, 14,* 6855-71.

Sengpiel, F., Stawinski, P. & Bonhoeffer, T. (1999). Influence of experience on orientation maps in cat visual cortex. *Nature Neuroscience, 2,* 727-32.

Sereno, A.B. & Maunsell, J.H.R. (1998). Shape selectivity in primate lateral intraparietal cortex. *Nature, 395,* 500-503.

Sermasi, E., Margotti, E., Cattaneo, A. & Domenici, L. (2000). Trk B signaling controls LTP but not LTD expression in the developing rat visual cortex. *European Journal of Neuroscience, 12,* 1411-9.

Shallice, M.R.C. (1988). Specialization within the semantic system. *Cognitive Neuropsychology, 5,* 133-142.

Shapley, R. & Perry, V.H. (1986). Cat and monkey retinal ganglion cells and their visual functional roles. *Trends in Neurosciences,* 229-235,

Sharma, S. C. (1975). Visual projection in surgically created "compound" tectum in adult goldfish. *Brain Research, 93,* 497-501.

Shatz, C.J., Ghosh, A., McConnell, S.K., Allendoerfer, K.L., Friauf, E., & Antonini, A. (1990). Pioneer neurons and target selection in cerebral cortical development. *Cold Spring Harbor Symposia on Quantitative Biology, 55,* 469-80.

Sheng, M. & Greenberg, M.E. (1990). The regulation and function of c-*fos* and other immediate early genes in the nervous system. *Neuron, 4,* 477-485.

Sherman, S.M., & Guillery, R.W. (1976). Behavioral studies of binocular competition in cats. *Vision Research, 16,* 1479-81.

Sherman, S.M., Guillery, R.W., Kaas, J.H., & Sanderson, K. (1974). Behavioral, electrophysiological, and morphological studies of binocular competition in the development of geniculocortical pathways of cats. *Journal of Comparative Neurology, 158,* 1-19.

Shimojo, S., Bauer, J., O'Connell, K.M., & Held, R. (1986). Pre-stereoptic binocular vision in infants. *Vision Research, 26,* 501-510.

Shinkman, P.G., Isley, M.R., & Rogers, D.C. (1985). Development of interocular relationships in visual cortex. In R.N. Aslin (Ed.), *Advances in neural and behavioral development, Vol. 1.* pp. 187-268.

Shipp, S. & Zeki, S. (1985). Segregation of pathways leading from area V2 to areas V4 and V5 of macaque monkey visual cortex. *Nature, 315,* 322-324.

Sillito, A.M. (1975). The contribution of inhibitory mechanisms to the receptive field properties of neurones in the striate cortex of the cat. *Journal of Physiology, London, 250,* 305-329.

Singer, W. (1985). Central control of developmental plasticity in the mammalian visual cortex. *Vision Research, 25,* 389-396.

Skottum, B.C., Bradley, A., Sclar, G., Ohzawa, I., & Freeman, R. (1987). The effects of contrast on visual orientation and spatial frequency discrimination: A comparison of single cells and behavior. *Journal of Neurophysiology, 57,* 773-786.

Smith, D.C. (1985). Behavioral and electrophysiological effects of visual deprivation and their reversal. In R.N. Aslin (Ed.), *Advances in neural and behavioral development, Vol. 1.* pp. 131-156.

Smith, D.C. & Holdefer, R.N. (1985). Binocular competitive interactions and recovery of visual acuity in long-term monocularly deprived cats. *Vision Research, 25,* 1783-94.

Smith, D.C., Spear, P.D., Kratz, K.E. (1978). Role of visual experience in postcritical-period reversal of effects of monocular deprivation in cat striate cortex. *Journal of Comparative Neurology, 178,* 313-28.

Smith, V.C., Lee, B.B., Pokorny, J., Martin, P.R. & Valberg, A. (1992). Responses of macaque ganglion cells to the relative phase of heterochromatically modulated lights. *Journal of Physiology, 458,* 191-221.

Snowden, R.J., Treue, S., Erickson, R.G., & Andersen, R.A. (1991). The response of area MT and V1 neurons to transparent motion. *Journal of Neuroscience, 11,* 2768-85.

Sokol, S. (1978). Measurement of infant acuity from pattern reversal evoked potentials. *Vision Research, 18,* 33-40.

Sparks, D.L. (1986). Translation of sensory signals into commands for control of saccadic eye movements: Role of primate superior colliculus. *Physiological Reviews, 66,* 118-171.

Spitzer, H., Desimone, R., & Moran, J. (1988). Increased attention enhances both behavioral and neuronal performance. *Science, 240,* 338-340.

Springer, A. D., Easter, S. S., & Agranoff, B.W. (1977). The role of the optic tectum in various visually mediated behaviors in goldfish. *Brain Research, 128,* 393-404.

Squire, L.R. (1986). Mechanisms of memory. *Science, 1986, 232,* 1612-9.

Stoerig, P. & Cowey, A. (1990). Blindsight and perceptual consciousness: Neuropsychological aspects of striate cortical function. In B. Gulyas, D. Ottolson, & P. E. Roland (Eds.), *Functional organization of the human visual cortex.* Oxford: Pergamon Press, pp. 181-193.

Stryker, M.P. & Harris, W.A. (1986). Binocular impulse blockade prevents the formation of ocular dominance columns in cat visual cortex. *Journal of Neuroscience, 6,* 2117-2133.

Stryker, M.P., Sherk, H., Leventhal, A.G., & Hirsch, H.V.B. (1978). Physiological consequences of effectively restricting early visual experience with oriented contours. *Journal of Neurophysiology, 41,* 896-909.

Stryker, M.P. & Strickland, S. L. (1984). Physiological segregation of ocular dominance columns depends upon the pattern of afferent electrical activity. *Investigative Ophthalmology and Visual Science, 25,* 278 (Suppl.).

Tanaka, K. (1993). Neuronal mechanisms of object recognition. *Science, 262,* 685-688.

Tanaka, K., Fukada, Y., & Saito, H. (1989). Underlying mechanisms of the response specificity of expansion/contraction and rotation cells in the dorsal part of the medial superior temporal area of the macaque monkey. *Journal of Neurophysiology, 62,* 642-656.

Tanaka, K., Hikosaka, K., Saito, H., Yukie, M., Fukada, Y., & Iwai, E. (1986). Analysis of local and wide-field movements in the superior temporal visual areas of the macaque monkey. *Journal of Neuroscience, 6,* 134-44.

Tao, X., Finkbeiner, S., Arnold, D.B., Shaywitz, A.J. & Greenberg, M.E. (1998). Ca^{2+} influx regulates BDNF transcription by a CREB family transcription factor-dependent mechanism. *Neuron, 20,* 709-726.

Teller, D. Y. (1979). The forced-choice preferential looking procedure: A psychophysical technique for use with human infants. *Infant Behavior and Development, 2,* 135-153.

Thorell, L.G., De Valois, R.L., & Albrecht, D.G. (1984). Spatial mapping of monkey V1 cells with pure color and luminance stimuli. *Vision Research, 24,* 751-769.

Tootell, R. B. H. & Taylor, J. B. (1995). Anatomical evidence for MT and additional cortical visual areas in humans. *Cerebral Cortex, 1,* 39-55.

Toyama et al. (1993). In vitro study of visual cortical development and plasticity. In Ono, T., Squire, L.R., Raichle, M.E., Perrett, D.I., Fukuda, M. (Eds.), *Brain mechanisms of perception and memory: From neuron to behavior.* New York: Oxford University Press, pp. 517-32.

Trachtenberg, J. T., Trepel, C. & Stryker, M. P. (2000). Rapid extragranular plasticity in the absence of thalamocortical plasticity in the developing primary visual cortex. *Science, 287,* 2029-2032.

Troyer, T. W., Krukowski, A. E., Priebe, N. J. & Miller, K. D. (1998). Contrast-invariant orientation tuning in cat visual cortex: thalamocortical input tuning and correlation-based intracortical connectivity. *Journal of Neuroscience, 18,* 5908-5927.

Ts'o, D.Y., Frostig, R.D., Lieke, E.E., & Grinvald, A. (1990). Functional organization of primate visual cortex revealed by high resolution optical imaging. *Science, 249,* 417-20.

Ts'o, D.Y. & Gilbert, C. D. (1988). The organization of chromatic and spatial interactions in the primate striate cortex. *Journal of Neuroscience, 8,* 1712-1727.

Tsumoto, T. (1993). Molecular mechanisms underlying long-term potentiation/depression in the developing visual cortex: An overview. In Ono, T., Squire, L.R., Raichle, M.E., Perrett, D.I., & Fukuda, M. (Eds.), *Brain mechanisms of perception and memory: From neuron to behavior.* New York: Oxford University Press, pp. 562-83.

Tsumoto, T. & Suda, K. (1979). Cross-depression: An electrophysiological manifestation of binocular competition in the developing visual cortex. *Brain Research, 168,* 190-4.

Udin, S.B. & Fawcett, J.W. (1988). Formation of topographic maps. *Annual Review of Neuroscience, 11,* 289-327.

Ungerleider, L.G. & Mishkin, M. (1982). Two cortical visual systems. In D.J. Ingle, R.J.W. Mansfield, & M.S. Goodale (Eds.), *The Analysis of Visual Behavior.* Cambridge, Mass: MIT Press, pp. 549-586.

Uusitalo, M. A., Jousmaki, V., & Hari, R. (1997). Activation trace lifetime of human cortical responses evoked by apparent visual motion. *Neuroscience Letters, 224,* 45-48.

Van Essen, D. C., Felleman, D. J., DeYoe, E. A., & Knierim, J. J. (1993). Probing the primate visual cortex: Pathways and perspectives. In B. Gulyas, D. Ottolson, & P. E.

Roland (Eds.), *Functional organization of the human visual cortex.* Oxford: Pergamon Press.

Vanni, S., Uusitalo, M. A., Kiesila, P. & Hari, R. (1997). Visual motion activates V5 in dyslexics. *Neuroreport, 8,* 1939-1942.

von Bonin, G. & Bailey, P. (1947). *The neocortex of Macaca mulatta.* Urbana: University of Illinois Press.

von der Heydt, R., Hanny, P., & Dursteler, M. R. (1980). The role of orientation disparity in stereoscopic perception and the development of binocular correspondence. In E. Grastyan & P. Molnar (Eds.), *Advances in physiological sciences, vol. 16: Sensory functions.* Oxford, England: Pergamon Press, pp. 461-470.

von Noorden, G.K. (1980). *Burian-Von Noorden's Binocular vision and ocular motility,* 2nd Edition. St. Louis, Missouri: Mosby.

von Noorden, G.K. (1981). New clinical aspects of stimulus deprivation amblyopia. *American Journal of Ophthalmology, 92,* 416-21.

Warren, W.H. & Hannon, J. (1988). Direction of self-motion is perceived from optic flow. *Nature, 336,* 162-163.

Warrington, E.K. & Shallice, T. (1984). Category specific semantic impairments. *Brain, 107,* 829-54.

Wässle, H. & Boycott, B.B. (1991). Functional architecture of the mammalian retina. *Physiological Review, 71,* 447-480.

Wässle, H., Yamashita, M., Greferath, U., Grunert, U. & Muller, F. (1991). The rod bipolar cell of the mammalian retina. *Visual Neuroscience, 7,* 99-112.

Watkins, J. C. (1994). The NMDA receptor concept: origins and development. In G. L. Collingridge & J. C. Watkins (Eds.), *The NMDA receptor,* 2nd Edition. Oxford: Oxford University Press, pp. 1-30.

Webster, M.J., Bachevalier, J. & Ungerleider, L.G. (1994) Connections of inferior temporal areas TEO and TE with parietal and frontal cortex in macaque monkeys. *Cerebral Cortex, 5,* 470-483.

Weiskrantz, L., Warrington, E.K., Sanders, M.D., Marshall, J. (1974). Visual capacity in the hemianopic field following a restricted occipital ablation. *Brain. 97,* 709-28.

Weliky, M., Katz, L.C. (1997). Disruption of orientation tuning in visual cortex by artificially correlated neuronal activity. Nature, 386, 680-5

Weliky, M. & Katz, L.C. (1999) Correlational structure of spontaneous neuronal activity in the developing lateral geniculate nucleus in vivo. *Science, 285,* 599-604.

Wernicke, C. (1874). *Der aphasische Symptomenkomplex.* Breslau: Cohn and Weigert.

Wheatstone, C. (1838). Contributions to the physiology of vision-Part first. On some remarkable and hitherto unobserved phenomena of binocular vision. *Philosophical Transactions of the Royal Society of London, 128,* 371-394.

White, A.J.R, Wilder, H.D., Goodchild, A.K., Sefton, A.J. & Martin, P.R. (1998). Segregation of receptive field properties in the lateral geniculate nucleus of a new-world monkey, the marmoset. *Callithrix jacchus. Journal of Neurophysiology, 80,* 2063-76.

Wiesel, T.N., & Hubel, D.H. (1963). Single-cell responses in striate cortex of kittens deprived of vision in one eye. *Journal of Neurophysiology, 26,* 1003-1017.

Wiesel, T.N. & Hubel, D.H. (1974). Ordered arrangement of orientation columns in monkeys lacking visual experience. *Journal of Comparative Neurology, 158,* 307-318.

Wild, H. M., Butler, S. R., Carden, D., & Kulikowski, J. J. (1985). Primate cortical area V4 important for colour constancy but not wavelength discrimination. *Nature, 313,* 133-135.

Williams, J. H. & Bliss, T. V. P. (1988). Induction but not maintenance of calcium-induced long-term potentiation in dentate gyrus and area CA1 of the hippocampal slice is blocked by nordihydroguaiaretic acid. *Neuroscience Letters, 88,* 81-85.

Winderickx, J., Lindsey, D.T., Sanocki, E., Teller, D.Y., Motulsky, A.G., & Deeb, S.S. (1992). Polymorphism in red photopigment underlies variation in colour matching. *Nature, 356,* 431-3

Wu, D.Y., Schneider, G.E., Silver, J., Poston, M., Jhaveri, S. (1998). A role for tectal midline glia in the unilateral containment of retinocollicular axons. *Journal of Neuroscience, 18,* 8344-55.

Yabuta, N.H. & Callaway, E.M. (1998). Functional streams and local connections of layer 4C neurons in primary visual cortex of the macaque monkey. *Journal of Neuroscience, 18,* 9489-99.

Young, A.W., Newcombe, F., Hellawell, D., De Hann, E. (1989). Implicit access to semantic information. *Brain and Cognition, 11,* 186-209.

Yuodelis, C. & Hendrickson, A. (1986). A qualitative and quantitative analysis of the human fovea during development. *Vision Research, 26,* 847-855.

Zafra, F., Castren, E., Thoenen, H. & Lindholm, D. (1991). Interplay between glutamate and gamma-aminobutyric acid transmitter systems in the physiological regulation of brain-derived neurotrophic factor and nerve growth factor synthesis in hippocampal neurons. *Proceedings of the National Academy of Science, USA, 88,* 10037-41.

Zeki, S.M. (1971). Cortical projections from two prestriate areas in the monkey. *Brain Research, 34,* 19-35.

Zeki, S.M. (1973). Colour coding in the rhesus monkey prestriate cortex. *Brain Research, 53,* 422-427.

Zeki, S.M. (1974). Functional organization of a visual area in the posterior bank of the superior temporal sulcus of the rhesus monkey. *Journal of Physiology, London, 236,* 549-573.

Zeki, S.M. (1978). Functional specialization in the visual cortex of the rhesus monkey. *Nature, 274,* 423-428.

Zeki, S.M. (1983a). Colour coding in the cerebral cortex: The reaction of cells in monkey visual cortex to wavelengths and colours. *Neuroscience, 9,* 741-765.

Zeki, S.M. (1983b). Colour coding in the cerebral cortex: The responses of wavelength-selective and colour-coded cells in monkey visual cortex to changes in wavelength composition. *Neuroscience, 9,* 767-781.

Zeki, S.M. (1990). A century of cerebral achromatopsia. *Brain, 113,* 1721-1777.

Zeki, S.M., Watson, J.D.G., Lueck, C.J., Friston, K.L., Kennard, C., & Frackowiak, R.S.J. (1991). A direct demonstration of functional specialization in human visual cortex. *Journal of Neuroscience, 11,* 641-649.

Zihl J, von Cramon D, Mai N, Schmid C. (1991). Disturbance of movement vision after bilateral posterior brain damage. Further evidence and follow up observations. *Brain, 114*, 2235-52.